Industrial Mathematics
A Course in Solving Real-World Problems

Industrial Mathematics
A Course in Solving Real-World Problems

Avner Friedman
Institute for Mathematics and its Applications
University of Minnesota

&

Walter Littman
University of Minnesota

Society for Industrial and Applied Mathematics
Philadelphia · 1994

Library of Congress Cataloging-in-Publication Data

Friedman, Avner
      Industrial mathematics : a course in solving real-world problems /
    Avner Friedman and Walter Littman
         p.   cm.
    Includes index.
    ISBN 0-89871-324-2
    1. Mathematics. 2. Mathematics—Industrial applications.
  I. Littman, Walter. II. Title
  QA37.2.F735 1994   94-20770
  515—dc20

Chapters 1 – 6 were adapted, with permission, from material that was presented in *Mathematics in Industrial Problems,* Parts 1 – 4, IMA Vol. Math. Appl., Vol. 16, © 1988, Springer-Verlag; Vol. 24 , © 1989, Springer-Verlag; Vol. 31, © 1990, Springer-Verlag; Vol. 38, © 1991, Springer-Verlag.

siam is a registered trademark.

Contents

Preface to the Student

You have had calculus of one and several variables. You have occasionally been asked to solve problems in physics and geometry, using what you have learned in calculus. Many of these problems, although a good illustration of the power of calculus, were solved by mathematicians and physicists several hundred years ago!

But what about today's real-world problems? Are calculus and "post" calculus (such as differential equations) playing an important role in research and development in industry? Are these mathematical tools indispensable for improving industrial products such as automobiles, airplanes, televisions, and cameras? Do they play a role in understanding air pollution, predicting weather and stock market trends, and building better computers and communication systems?

This book was written to convince you, by examples, that the answer to all the above questions is YES! Indeed, each chapter presents one important problem that arises in today's industry, and then gets down to studying the problem by mathematical analysis and computation. Numerical work and mathematical analysis go hand in hand and build on each other toward a better understanding of the problems under study.

We assume only that you have taken calculus of one and several variables. However, in each chapter we will also need to develop a little new mathematics to tackle the industrial problem. When you have gone through this book, you will have learned many new ideas and methods from ordinary and partial differential equations and from integral equations and control theory.

Our goal is to impart to you the excitement and usefulness of mathematics as a tool in solving real-world problems. We have chosen problems that on one hand are of great importance to industry, and on the other hand can be studied by students whose background is only calculus. We emphasize that this book is not a substitute for texts on ordinary differential equations, partial differential equations, integral equations, or other similar standard courses taught at the undergraduate junior/senior level or at the graduate level. Here we develop ideas and methods from the above fields rather casually, just as needed for solving the industrial problems. In keeping up the goal of placing you at the cutting edge of the industrial mathematics research, we have selected, in each chapter, just a few problems, as they are relevant to the content of the chapter; these are not meant to be just drill exercises. We have included at the end of each chapter references to the problem under study.

Have fun!

Introduction

Industrial mathematics is a fast growing field within the mathematical sciences. It is characterized by the origin of the problems that it engages; they all come from industry: research and development, finances, and communications. Naturally, industrial mathematics may borrow from a variety of mathematical disciplines, such as partial differential equations, dynamical systems, control theory, probability and statistics, and discrete mathematics. The common feature running through this enterprise is the goal of gaining better understanding of industrial models and processes through mathematical ideas and computations.

Historically, applications have been the driving force in the sciences and in mathematics. However, applications can also serve to motivate the teaching of mathematics in school! In this book we have undertaken the approach of presenting real industrial problems and their mathematical modeling as a motivation for developing mathematical methods that are needed for solving the problems. We have selected the problems from the collection of six volumes authored by Avner Friedman.[1]

The problems selected are accessible to undergraduate students who have already taken what in many colleges and universities constitutes the first two-year basic calculus sequence. Such a sequence includes functions of one and several variables and a rudimentary knowledge of the following topics: ordinary differential equations, linear algebra, infinite series, and vector analysis. In addition, a working knowledge of Fortran, Pascal, or C language is required.

By the end of this course, the students will have mastered some basic material in theory and computations of ordinary and partial differential equations, integral equations, calculus of variations, and control theory. However, this material by no means represents the full scope of the standard

---

[1] A. Friedman, *Mathematics in Industrial Problems*, IMA Vol. Math. Appl., Springer-Verlag, New York, Berlin, Vol. 16, 1988; Vol. 24, 1989; Vol. 31, 1990; Vol. 38, 1991; Vol. 49, 1992; Vol. 57, 1993.

theory that is taught in one-semester or one-year courses on each of these topics. Thus the present course should not be considered a substitute for the traditional courses in differential and integral equations, but rather be viewed as motivation for taking such courses.

The book closely follows a one-year course taught at the University of Minnesota during 1992–1993 by Walter Littman and Bernardo Cockburn. A SUN workstation was provided for use by the students in that course. Some computations were also performed on Apollo workstations or whatever systems the students had available in their own departments.

The bulk of the students' effort (at least if measured in time spent) was in performing the computational homework. As a matter of fact, a computational project such as that mentioned in Note 2 of §6.7 might very well take the place of a final examination.

We have found that the academic diversity of the students was a great asset. There were students from mathematics, physics, computer science, and various engineering departments present, each having their own points of view to present. For example, if a student has had occasion to use the conjugate gradiant method in another course, he or she might be encouraged to present it to the class. While tackling the industrial problems, most students found the computing challenges most exciting, while others found the greatest motivation in the theory.

One attitude to encourage is that the class is one research organization attacking new problems. Thus, it makes sense to divide the problems among various students or groups of students. Different students might be asked to do the same computation with different sets of numerical data and their results may be compared. To facilitate this, graphing of results (including three-dimensional graphs) should be encouraged. The more theoretically motivated students might be assigned the more challenging theoretical problems and variations of these.

A very important aspect of the course is the quality of the teaching assistant who also grades the homework. Since his or her job also entails helping students master the intricacies of the computer systems, this can be a time-consuming job.

We hope this book will convince students that mathematics is useful and very much alive in real-life problems. We live in an increasingly technological society, where mathematical skills are required at many levels, and we hope the students will learn to appreciate and use mathematics in whatever career they choose.

We would like to thank Peter Castro, David Ross, and John Spence from Eastman Kodak for proposing the problems presented in Chapters 1, 4, and 6–7, respectively; to David Chock from Ford Motor Company for proposing the problem studied in Chapter 2; to Dean Gerber from IBM Yorktown Heights for proposing the problem presented in Chapter 3, and to Ed Bissett from General Motors Research and Development Center for proposing the problem described in Chapter 5. We benefited from many discussions we have had with them.

Thanks are also due to Bernardo Cockburn for contributing §§2.8, 4.6, and 6.8, dealing with numerical methods. He also devised codes for a number of computational problems, which will be available on the Internet. Tim Brule was the teaching assistant; he and students Shijie Deng, Jens Henrickson, Eric Jerkins, John Liljeberg, Yuan Lou, Dan Sargent, Jan Sefcik, and Jonathan Thom have provided numerical results and graphs, some of which are presented in the text. Dan Drucker from Wayne State University sent us a long list of corrections. We thank them all for their contributions.

We wish to thank Patricia V. Brick for typing the manuscript.

We would like thank our wives, Lynn and Florence, for their support and encouragement.

Finally it is a pleasure to acknowledge the support of the National Science Foundation grant USE–9250106 to the development of this project.

Chapter 1

Crystal Precipitation

1.1. The road ahead—Some helpful hints to the student

Since this is the first chapter, it is appropriate to say a few words about the general structure of these chapters and some advice on how to approach them.

You will note that §1.2, called "Background," is a short physical, non-mathematical description of the problem, which in this chapter is to gain an understanding of how certain crystal grains grow in a solution.

In §1.3, called "The model," you are confronted by what may seem like a daunting collection of equations and formulas. The point here is to try to see some *logical structure* in the collection of equations and formulas derived by chemists and physicists by a variety of methods. This happens in (1.7) and (1.8). It is here that we note that the complicated equations written earlier really have a very simple form: a system of ordinary differential equations with initial conditions—one of the most useful tools in applied mathematics.

Now if we are going to use a tool, we had better learn *how* to use it. That is the purpose of §§1.4–1.7, where some basic facts about differential equations are discussed. Note the word "BASIC" after the title of a section. This indicates that we are discussing a *basic tool* that may be used over and over again.

Since we do not only want to understand the *theoretical* aspects of our problem, but would also like to be able to *compute* solutions to problems, it is important to be able to *compute* solutions to differential equations *numerically*, since with complicated differential equations *explicit* solutions are usually not available the way they are in elementary courses. One such method is discussed in §1.7.

In §1.8 we apply our newly acquired tools (both the theoretical and the computational tools) to gain insight into our problem. However, we begin this section by looking at a very special situation, which can be described by a *single* differential equation.

The idea here is that before attacking a difficult problem, we should attack a "warm-up" problem—one that is easier but still contains the essential features of the original problem. The solution of the warm-up problem, it is hoped, will shed some light and give us confidence to attack the full problem.

In §1.11 a much more efficient numerical method for solving ordinary differential equations (the "Runge–Kutta method") is introduced, while in §1.13, buttressed by the experience and confidence gained in §1.8, we attack the original model—described by a *system* of differential equations.

1.2. Background

A photographic *emulsion* is a suspension of small particles in aqueous gelatin. A *crystal* is a three-dimensional atomic or molecular structure consisting of periodically repeated unit cells. Crystal particles often have planar external faces. A modern photographic film consists of several emulsion layers. One of these layers is shown in Fig. 1.1; the suspended particles are silver halide crystal grains. The most common form of silver halide used is silver bromide, "AgBr."

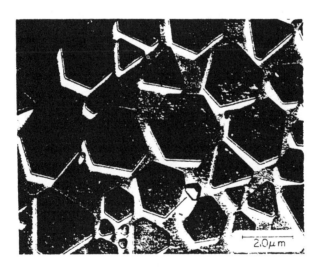

FIG. 1.1. *Silver halide grains.*[1]

When you shoot a picture, the shutter in the camera opens and the film is exposed to light. A stream of photons penetrates the emulsion layers of the film, and some enter the halide crystal grains, causing silver ions to get "trapped" at some locations on the grains' boundaries. When the film is

[1]A. Friedman, *Mathematics in Industrial Problems*, IMA Vol. Math. Appl., Vol. 16, Springer-Verlag, New York, 1988, Fig. 4.4, p. 33. ©1988, Springer-Verlag.

developed, the silver halide grains with trapped silver ions show up as black spots on the negative in a black and white film. In this way, the darkened areas of the film correspond to light areas in front of the camera. The process of "printing" or enlarging then produces a positive image where light and dark are again interchanged.

"Graininess" in a photograph is a cluster phenomenon of specks appearing in what should be a solidly colored area. It is caused by the innate graininess of the process: as described above, the picture is produced by discrete silver halide grains. The larger the grains in the film, the grainier the pictures will be. It may seem obvious what to do: use small grains. But there is a catch. Small grains do not capture light as efficiently as large ones. Films that have small grains are "slow": they require more time to capture a scene than those with large grains, which are "fast." (This difference is noted on film packages, e.g., you have a choice between 400 speed and 200 speed when you buy film in the drug store.) A person can only hold a camera still for about 1/30 of a second, and this may not be enough exposure time for darker scenes if one uses small-grain film. Thus there is a trade-off between film speed and graininess: we tolerate a bit of graininess to increase speed while we sacrifice speed to reduce graininess. Even in a single film, we may want some large grains, so that we do not miss the features in the shadows entirely, and some small grains, so that the features in bright regions are not grainy. To make this trade-off, *one must be able to control the sizes and distribution of silver halide crystal grains* in the film.

How are silver halide crystal grains put in photographic film? They are first mixed into a wet gelatin emulsion layer. This wet emulsion is then dried, so that the film is actually dry. Now as we have seen, it is desirable to have silver halide grains of several sizes in prescribed proportions. So the question boils down to this: how can we manufacture grains of a *given* size? Once we know how to obtain grains of a given size, we can then keep the various sized grains in different "pots" and when needed mix them in the proper proportions into gelatinous layers to make a film with the desired balance of graininess.

The method of manufacturing silver halide grains of a given (approximate) size is based on a process called "Ostwald ripening."

The process begins with a dry mixture of small size grains of various sizes that have been obtained by a "precipitation" process from a solution. For example, if we mix *solutions* of silver nitrate ($AgNO_3$) and potassium bromide (KBr), the following reaction takes place:

$$AgNO_3 \text{ (solution)} + KBr \text{ (solution)} \rightarrow AgBr \text{ (solid)} + KNO_3 \text{ (solution)}.$$

$AgBr$ thus *precipitates* out of the solution in the form of crystals of various sizes. This first step is very easy because it is not required that the grains be nearly equal in size. Next, this mixture of grains is added to a certain solvent and kept mixed. The phenomenon of Ostwald ripening is based on the following fact: If one allows this process to continue for a "very very long time," either all crystal grains will dissolve into solution, or all the grains in the solution become the same size. So the objective is to arrange things (by choosing the proper concentrations, for example) so that the second possibility occurs.

Now since the manufacturers of film are also motivated by the economic efficiency of the process, they do not like to wait a "very very long time." Thus they stop the process after only a "long period of time" (say minutes or hours) by taking the liquid out of the solution, and content themselves with pots of grains of "approximately equal sizes."

It is interesting to note that even after the photographic film has been made, the silver halide crystals in the film continue to undergo this "Ostwald ripening" effect, but at a very slow rate, and the quality of the film is not appreciably affected for several years.

Incidentally, color film contains, in addition to silver halide crystals, oil droplets, as will be discussed in Chapter 4. These oil droplets also undergo the same process of Ostwald ripening and the mathematical analysis of this chapter applies to them also. The only difference is that the initial formation of the oil droplets is accomplished not by precipitation from a solution, but by the action of a blender.

Generally, chemical engineers tend to use empirical equations (i.e., equations derived experimentally) to describe the evolution of the crystal size distribution. This presents difficulties in changing from one set of process conditions to another, since new empirically developed equations may have to be established for a new situation. Additionally, there is difficulty in predicting which changes in the process will improve results. In the drive to produce the highest quality product at the least cost it becomes ever more important to have the understanding that can come from the mathematical models. One commonly used model is described below.

Now, all mathematical models of physical phenomena make certain simplifying assumptions. For example, Galileo's law of falling weights neglects air resistance. The model for Ostwald ripening described below is no exception. However, to make the treatment possible at the undergraduate level we have made one additional simplifying assumption: We assume that *all crystal grains have the same shape* and that they differ only in size. For example, they may all be boxes with edges $(\lambda a, \lambda b, \lambda c)$ where a, b, c are fixed positive numbers and λ is a positive variable. For definiteness, we assume that all the grains are cubes with variable diameter and variable orientation.

1.3. The model

As we have mentioned, the evolution of crystals in a fluid is a phenomenon, called *Ostwald ripening*, in which small crystals tend to dissolve while larger crystals tend to approach a uniform size.

Given a volume of fluid containing an amount of dissolved matter (solute), there will be in equilibrium a saturation concentration c^*, which is the maximum solute per unit volume of fluid that the system can hold. If the actual concentration $c(t)$ exceeds c^*, then the excess precipitates out in solid form, i.e., in crystal form. Actually, to cause precipitation, $c(t)$ must be larger than a quantity c_L ($c_L > c^*$), which also depends on the size of the grain, as will be explained below. The variable t represents time (measured in seconds).

We denote the edge of crystal grains by L (recall that they are all cubes). We assume that the distribution of grains of edge L is roughly uniform throughout the solution. Then c_L is given by

$$c_L = c^* e^{\Gamma/L} \qquad \text{(Gibbs–Thomson relation)},$$

where Γ is a physical quantity that depends on the shape of the crystals, on its material properties, and on the temperature (which is assumed fixed). If $c(t) > c_L$, then material will come out of the solution and deposit onto the crystals characterized by L, and if $c(t) < c_L$ then material will dissolve from the crystals. Set

(1.1) $$L^*(t) = \frac{\Gamma}{\log \frac{c(t)}{c^*}}.$$

Here "log" means *natural log*, i.e., logarithm to base $e = 2.718\ldots$.

According to semi-empirical law, the crystal size L will grow or dissolve at the rate

(1.2) $$\frac{dL}{dt} = G(L, c(t)),$$

where

(1.3) $$G(L, c(t)) = \begin{cases} k_g(c(t) - c^* e^{\Gamma/L})^g & \text{if } L > L^*(t), \\ -k_d(c^* e^{\Gamma/L} - c(t))^d & \text{if } L < L^*(t) \end{cases}$$

and where k_g, k_d, g, d are positive constants,

(1.4) $$1 \le g \le 2, \qquad 1 \le d \le 2.$$

Observe that, by (1.2), (1.3),

$$\text{if } c(t) > c_L \text{ (or } L > L^*(t)) \text{ then } \frac{dL}{dt} > 0, \quad \text{i.e., the crystal grows,}$$

if $c(t) < c_L$ (or $L < L^*(t)$) then $\dfrac{dL}{dt} < 0$,   i.e., the crystal shrinks.

Let us assume that initially there are N different sizes of crystals, characterized by sizes $L = x_j^*$ in numbers μ_j^* per unit volume, where

$$0 < x_1^* < x_2^* < \cdots < x_N^*.$$

These sizes will then evolve in time to $x_1(t), \ldots, x_N(t)$, according to (1.2),

(1.5) $$\frac{dx_j}{dt} = G(x_j, c(t)).$$

The concentration $c(t)$ of the solute at time t is given by

(1.6) $$c(t) = c_0 + \rho k_v \sum_{j=1}^{N} \mu_j^* (x_j^*)^3 - \rho k_v \sum_{j=1}^{N} \mu_j^* x_j^3(t),$$

where c_0 is the initial concentration, k_v is a geometric parameter connecting L^3 to the crystal volume (in the case of cubic crystals, $k_v = 1$), and ρ is the mass density of the solid phase.

If we substitute $c(t)$ from (1.6) into (1.5), we obtain a system of differential equations

(1.7) $$\frac{dx_j}{dt} = G_j(x_1, \ldots, x_N) \qquad (j = 1, \ldots, N).$$

We also have initial conditions

(1.8) $$x_j(0) = x_j^*.$$

Set

(1.9) $$\mu_j = \rho k_v \mu_j^*, \qquad c_1 = c_0 + \sum_{j=1}^{N} \mu_j (x_j^*)^3.$$

Note that c_1 represents the total amount of silver halide per unit volume in either crystal or solution form.

In the case where $N = 1$,

(1.10) $$\frac{dx}{dt} = G(x), \qquad x(0) = x^*,$$

where

(1.11) $$G(x) = \begin{cases} k_g(c_1 - \mu x^3 - c^* e^{\Gamma/x})^g & \text{if} \quad c_1 - \mu x^3 > c^* e^{\Gamma/x} \\ -k_d(c^* e^{\Gamma/x} - (c_1 - \mu x^3))^d & \text{if} \quad c^* e^{\Gamma/x} > c_1 - \mu x^3. \end{cases}$$

Typical physical constants are

$$c^* = 4 \times 10^{-6}\text{kmol }/\text{m}^3, \qquad \Gamma = 4 \times 10^{-9}\text{m},$$
$$\rho = 6473\text{kg/m}^3, \quad k_g = k_d = 5 \times 10^{-2}, \qquad d = 1$$
$$\text{and } g = 1 \quad \text{or} \quad g = 2, \quad \text{and} \quad c_0 = 1.05c^*,$$
$$x^* = 10^{-7}\text{m} \quad \text{or} \quad 10^{-8}\text{m} \quad \text{and} \quad \mu = 10^{19} \quad \text{or} \quad 10^{16}.$$

Note on "kmol". kmol stands for "kilomole" or 1000 moles. A kilomole of silver bromide is 188 kilograms of silver bromide, for example; this number is simply the sum of the atomic weights of Ag and Br. In calculations, one should use this number to convert all concentrations to kilograms per cubic meter (kg/m^3).

Equation (1.10) is an ordinary differential equation with initial condition, while (1.7) and (1.8) present a system of ordinary differential equations with initial conditions. To be able to discuss these intelligently, we must study such equations in some detail.

1.4. Some facts about differential equations (BASIC)

In this section we review some basic facts about *ordinary* differential equations. An ordinary differential equation is a relationship between an unknown function $y(t)$ of *one* independent variable and some of its derivatives, as well as the independent variable t. Usually this relationship can be expressed in the form

$$y^{(n)}(t) = f(t, y(t), \ y'(t), \ldots, y^{(n-1)}(t)).$$

Here $y', \ldots y^{(n-1)}, y^{(n)}$ represent the first through the nth derivative of $y(t)$. "n" here is called the *order* of the equation and represents the highest-order derivative that appears in the equation.

In this section we will be mainly interested in *first-order equations* and also in systems of first-order equations.

Suppose the function $f(t, y)$ is given as a continuous function in a rectangle R in the (t, y) plane, having its sides parallel to the t and y axes. The first-order equation

(1.12) $$\frac{dy}{dt} = f(t, y)$$

then assigns to every point (t, y) in R a slope dy/dt of a possible solution going through that point. Another way to look at this is that the equation prescribes a "direction field" (see Fig. 1.2), i.e., through every point in R a direction (or unit vector) is assigned. The unit vector at the point (t, y) has components

$$\left(\frac{1}{\sqrt{1 + (f(t, y))^2}}, \quad \frac{f(t, y)}{\sqrt{1 + (f(t, y))^2}} \right)$$

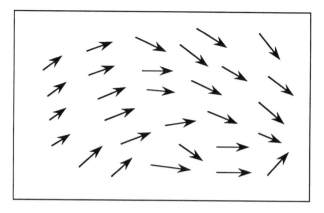

FIG. 1.2.

(why?). The problem of solving (1.12) may be thought of as finding curves whose tangent vectors are parallel to the vectors of the "direction field." It is intuitively clear that there are many such curves, and it seems reasonable to expect the curve to be uniquely determined if we prescribe a point (t_0, y_0) that the curve should pass through. That this is indeed the case (under appropriate assumptions on $f(t, x)$) is one of the basic facts of the theory.

In looking for solutions to a mathematical problem two questions are usually foremost: one, does a solution exist, and, two, is the solution unique. In discussing differential equations—which play a prominent role in describing physical phenomena—both questions are important. Indeed, suppose we model a physical system (such as, for example, a swinging pendulum, or a vibrating spring) by a differential equation. If a solution to the equation (together with "initial condition") does not exist, then clearly something is wrong with the model, since the physical system obviously does exist and function in the real world! Thus the question of existence is very important for the validation of the model. We may wish to predict the behavior of the physical system by solving the differential equation (together with "initial condition"). If the solution of the equation is not uniquely determined by the initial data, how do we know that the mathematical solution that we can compute is really the solution that "nature" picks? Thus the question of uniqueness is also important.

We now state the well-known theorems on existence and uniqueness for the equation

$$(1.13) \qquad\qquad y' = f(t, y)$$

with the initial condition

$$(1.14) \qquad\qquad y(t_0) = y_0.$$

We will consider this "initial value problem" in a rectangle R given by

$$|t - t_0| \leq a, \qquad |y - y_0| \leq b.$$

Suppose that f and $\partial f/\partial y$ are continuous functions in R. Since R is a closed bounded set, this implies that f and $\partial f/\partial y$ are actually *bounded* in R:

$$|f(t,y)| \leq M, \qquad \left|\frac{\partial f}{\partial y}(t,y)\right| \leq K \quad \text{in } R.$$

By the law of the mean applied to $f(t,y)$ as a function of y, it follows that

$$f(t,y_2) - f(t,y_1) = \frac{\partial f}{\partial y}(t,y^*)(y_2 - y_1)$$

for some value y^* between y_1 and y_2, and hence

(1.15) $$|f(t,y_2) - f(t,y_1)| \leq K|y_2 - y_1|$$

as long as (t,y_1) and (t,y_2) are points in R.

A function $f(t,y)$ satisfying an inequality of the type (1.15) is said to satisfy a *Lipschitz condition* (with Lipschitz constant K) in R.

THEOREM 1.4.1. *Suppose $f(t,y)$ is defined in the rectangle*

$$R: \quad |t - t_0| \leq a, \qquad |y - y_0| \leq b,$$

that f is continuous in R and satisfies $|f(t,y)| \leq M$ and the Lipschitz condition (1.15) there. Further assume that $b \geq aM$. Then there exists exactly one solution $y(t)$ to the initial value problem (1.13), (1.14) with $|t - t_0| < a$, and with graph lying in R.

The solution $y(t)$ will, of course, depend on the value y_0 (or briefly, call it η now) at $t = t_0$, although the dependence on $y_0 \equiv \eta$ is not indicated by our notation. As a by-product of the uniqueness proof, it can indeed be shown that the solution $y(t) = y(t,\eta)$ (indicating its dependence on the assigned initial value η prescribed at $t = t_0$) is indeed *a continuous function of both t and η*, as well as *a continuously differentiable function of t*.

We will now give a proof of the *uniqueness* part of Theorem 1.4.1.

Suppose $y_1(t)$ and $y_2(t)$ are solutions of the initial value problem (1.13), (1.14) with graphs lying in R. Let $v(t) = y_2(t) - y_1(t)$. Then $v(t_0) = 0$ and $v'(t) = f(t, y_2(t)) - f(t, y_1(t))$; hence

$$v(t) = \int_{t_0}^{t} (f(\tau, y_2(\tau)) - f(\tau, y_1(\tau))d\tau,$$

so that

$$|v(t)| \leq K \int_{t_0}^{t} |y_2(\tau) - y_1(\tau)| d\tau = K \int_{t_0}^{t} |v(\tau)| d\tau.$$

Now for t in the interval $I_h = [t_0, t_0 + h]$

$$\max_{I_h} |v(t)| \leq K \int_{t_0}^{t_0+h} |v(t)| d\tau \leq Kh \max_{I_h} |v(t)|.$$

Hence if $K|h| \leq \frac{1}{2}$ we see that $v(t)$ must vanish identically in I_h. Similarly, $v(t)$ must vanish for $t_0 - h \leq t \leq t_0$. If necessary, we can continue this procedure by starting at the point $t_1 = t_0 + h$ or $t_1 = t_0 - h$ and concluding that $v(t)$ must vanish in a series of neighboring intervals of size $h = 1/2K$ until the whole interval $|t - t_0| < a$ is covered.

1.5. Picard's method of successive approximations (BASIC)

Next we describe the method of successive approximation, which can be used to construct an approximation to the solution of the initial value problem, and perhaps more importantly, is the basis for the existence proof.

First we observe that the *initial value problem* (1.13), (1.14) is equivalent to the integral equation

$$y(t) = y_0 + \int_{t_0}^{t} f(\tau, y(\tau)) d\tau.$$

We now define a sequence of functions iteratively as follows:

$$y_0(t) \equiv y_0,$$

$$y_n(t) = y_0 + \int_{t_0}^{t} f(\tau, y_{n-1}(\tau)) d\tau.$$

Under the assumptions of Theorem 1.4.1, it can be shown that this sequence of continuous functions converges to a continuous limit function $y(t)$ (in the whole interval $|t - t_0| < a$), which is a solution of the integral equation; hence it is also a solution of the original initial value problem.

This method can also be used as a basis for numerical computation, provided the integrations can be carried out simply. If that is not the case and numerical integrations must be performed, it is usually preferable to use other numerical methods (which we describe in §§1.7 and 1.11).

What we are doing generally is drawing a straight line starting at the point
$P_0 = (a, \eta)$ with slope $f(a, \eta)$ and stopping at the point P_1 having coordinates
$(a + h, \eta + hf(a, \eta)) \equiv (t_1, y_1)$. Then, starting at P_1 we draw another straight
line with slope $f(t_1, y_1)$. We continue this procedure, obtaining a polygon with
vertices $P_0, P_1, P_2, \ldots, P_n$, where the point P_k has coordinates (t_k, y_k) given
recursively by

$$t_0 = a, \qquad y_0 = \eta,$$

$$t_{k+1} = t_k + h, \quad y_{k+1} = y_k + hf(t_k, y_k), \qquad k = 0, 1, \ldots, n - 1.$$

Because of the recursive nature of this method, a computational procedure
can easily be devised. Of course, the smaller the value for h is taken, the
more accurate the procedure becomes. Indeed, it can be shown that, under
the assumptions made in the previous section for existence and uniqueness, as
$h \to 0$ the polygonal curve (which represents a continuous function) approaches
another continuous function $y(x)$, which is a solution to the initial value
problem. The proof of this fact is another proof of the existence of a solution
to the initial value problem.

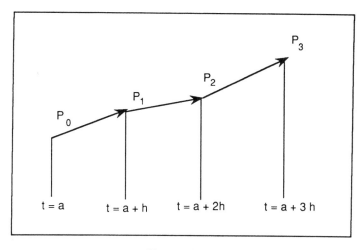

FIG. 1.3.

The method works just as well for systems. Consider, for example, the
initial value problem

$$\frac{du}{dt} = p(t, u, v),$$

$$\frac{dv}{dt} = q(t, u, v),$$

$$u(0) = \mu,$$

$$v(0) = \nu.$$

We introduce the column vectors

$$\begin{pmatrix} u \\ v \end{pmatrix} = \mathbf{y}, \quad \begin{pmatrix} \mu \\ \nu \end{pmatrix} = \boldsymbol{\eta}, \quad \text{and} \quad \begin{pmatrix} p(t, u, v) \\ q(t, u, v) \end{pmatrix} = \mathbf{f}(t, \mathbf{y}).$$

Then, in terms of $\mathbf{y}$ and $\mathbf{f}$, the initial value problem can be expressed as

$$\frac{d\mathbf{y}}{dt} = \mathbf{f}(t, \mathbf{y}), \qquad \mathbf{y}(0) = \boldsymbol{\eta}.$$

The Euler method can now be described vectorially as follows:

$$t_0 = 0, \qquad \mathbf{y}_0 = \boldsymbol{\eta},$$

$$t_{k+1} = t_k + h, \qquad \mathbf{y}_{k+1} = \mathbf{y}_k + h\mathbf{f}(t_k, \mathbf{y}_k) .$$

The last relationship can be expressed in component as follows:

$$u_{k+1} = u_k + hp(t_k, u_k, v_k),$$

$$v_{k+1} = v_k + hq(t_k, u_k, v_k),$$

while $\mathbf{y}_0 = \boldsymbol{\eta}$ becomes $u_0 = \mu$, $v_0 = \nu$.

PROBLEM 1.7.1. Compute numerically the solutions to the differential equation with initial condition:
$$\frac{dx}{dt} = 2x - 2t^2 - 3, \qquad x = 2 \quad \text{when } t = 0$$

for the interval $0 \leq t \leq 2$, and compare with the explicit solution found in Problem 1.6.1.

PROBLEM 1.7.2. Do the same for the differential system described in Problem 1.6.2 for the same interval.

1.8. Crystals of single size

Consider the situation when all crystal grains have initially the same size x^*. The crystals will evolve according to (1.10), (1.11). The points x where G changes sign, that is, where

$$(1.18) \qquad \mu x^3 + c^* e^{\Gamma/x} = c_1 \qquad (= c_0 + \mu(x^*)^3),$$

will play a crucial role.

THEOREM 1.8.1. *There exist at most two positive solutions* ξ_1, ξ_2 ($\xi_1 \leq \xi_2$) *of* (1.18).

Indeed, this follows from the fact that the function $f(x) = \mu x^3 + c^* e^{\Gamma/x}$ has a positive second derivative for $x > 0$, and $f(x) \to \infty$ if $x \to 0$ or if $x \to \infty$.

PROBLEM 1.8.1. Provide the details of the proof of Theorem 1.8.1.

Equation (1.18) may have no positive solutions, but for definiteness we concentrate on the case where there are two distinct positive solutions.

THEOREM 1.8.2. *If $x^* = \xi_1$ or $x^* = \xi_2$ then the solution of (1.10) is* $x(t) \equiv x^*$.

Indeed, $x(t) \equiv x^*$ is a solution and, by uniqueness, it is the only solution.

PROBLEM 1.8.2. Why does uniqueness hold for (1.10)?

Observe that

$$\mu x^3 + c^* e^{\Gamma/x} < c_1 \quad \text{if } \xi_1 < x < \xi_2,$$

$$\mu x^3 + c^* e^{\Gamma/x} > c_1 \quad \text{if } x < \xi_1 \text{ or if } x > \xi_2.$$

This implies that

$$G(x) > 0 \quad \text{if } \xi_1 < x < \xi_2,$$

$$G(x) < 0 \quad \text{if } x < \xi_1 \text{ or if } x > \xi_2.$$

Consequently

$$\frac{dx}{dt} > 0 \quad \text{if } \xi_1 < x < \xi_2,$$

$$\frac{dx}{dt} < 0 \quad \text{if } x < \xi_1 \text{ or if } x > \xi_2 .$$

In particular,

if initially $\quad x^* > \xi_2 \quad$ then the crystal

size will decrease for all $\ t > 0.$

The size $x(t)$ will remain $> \xi_2$, for otherwise we obtain $x(t_0) = \xi_2$ for some $t_0 > 0$ and this is a contradiction to the uniqueness.

PROBLEM 1.8.3. Prove Theorems 1.8.3–1.8.5.

THEOREM 1.8.3. *If $x^* > \xi_2$ then* $\lim_{t \to \infty} x(t) = \xi_2$.

Hint. If $\lim_{t \to \infty} x(t) = \overline{\xi}_2 > \xi_2$ then $x'(t) \leq -c$ for all $t > 0$, where c is a positive constant.

THEOREM 1.8.4. *If $0 < x^* < \xi_1$ then $x(t)$ is strictly decreasing for some time interval $0 \leq t \leq t_0$ and $x(t_0) = 0$.*

THEOREM 1.8.5. *If $\xi_1 < x^* < \xi_2$ then $x(t)$ strictly increases for all positive times, and* $\lim_{t \to \infty} x(t) = \xi_2$.

Note that Theorem 1.8.4 asserts that the crystal's size will decrease to zero in finite time t_0; thereafter the crystal is entirely dissolved in the solution (there is no differential equation for $x = 0$).

1.9. Remarks on Theorems 1.8.3–1.8.5

To establish the existence and uniqueness of the differential equation $dx/dt = G(x), x(0) = x^*$, we must appeal to the basic existence theorem (Theorem 1.6.1) with $f(t,y)$ satisfying a Lipschitz condition (why?) (See the remark following Theorem 1.6.1.) This theorem will suffice in Theorems 1.8.3 and 1.8.5 because there we can limit ourselves to values of x restricted to intervals of the type

$$0 < a \le x \le b,$$

in which G is bounded and satisfies a Lipschitz condition with a fixed constant K. In Theorem 1.8.4, on the other hand, since $x(t) \to 0$ (or so we hope) we must allow x to vary in an interval of the type

$$0 < x \le b,$$

in which G is not bounded, because of the term $e^{\Gamma/x}$ in the definition of $G(x)$. One way to get around that difficulty, is to write the differential equation in the form

$$\frac{dt}{dx} = 1/G(x), \qquad t(x^*) = 0,$$

and to consider x as the independent variable. We then obtain

$$t(x) = \int_{x^*}^{x} \frac{dx'}{G(x')}.$$

1.10. Reminder on Newton's method

Newton's method is an iterative method for finding roots (solutions) of the equation $f(x) = 0$. If this equation is written in the form

$$x = x - \frac{f(x)}{f'(x)},$$

or $x = T(x)$ where $T(x) = x - f(x)/f'(x)$, an iterative procedure (i.e., one of successive approximation)

$$x_1 = T(x_0), \qquad x_2 = T(x_1), \ldots, x_n = T(x_{n-1}), \ldots$$

is reasonable. Under the right conditions, x_n will converge to a root of $f(x) = 0$. Much depends on an intelligent choice of x_0. If there are several roots, different choices of x_0 may result in approximating sequences to the various roots. Therefore careful choice of x_0 is important!

PROBLEM 1.10.1. Solve $x = \cos x$ using Newton's method.

In Problems 1.10.2–1.10.4 you will be asked to verify Theorems 1.8.3–1.8.5 numerically. To facilitate this we will choose the constants g and d in (1.11) to be one ($g = d = 1$), thus yielding the simpler equation

(E)
$$\frac{dx}{dt} = k(c_1 - \mu x^3 - c^* e^{\Gamma/x}),$$

where we have taken $k_g = k_d = k$. To make the numbers somewhat more manageable we convert to new units: We use microns (μm) for length and picograms (pg) for mass:

$$1 \text{ micron} = 10^{-6} \text{ meters},$$

$$1 \text{ picogram} = 10^{-12} \text{ grams} = 10^{-15} kg,$$

$$1 \frac{\text{gram}}{cm^3} = 1 \frac{pg}{\text{microns}^3}.$$

We keep the time units as seconds.

In each of the following problems you will first have to find the two roots ξ_1 and ξ_2 of the equation

(1.19)
$$\mu x^3 + c^* e^{\Gamma/x} = c_1 \qquad (= c_0 + \mu(x^*)^3)$$

using Newton's method. Since ξ_1 and ξ_2 may be very close together, care will have to be taken to obtain both roots. You should choose your initial guesses in Newton's method with care. Try a rough graph of the left-hand side of (1.19) if necessary.

PROBLEM 1.10.2. Solve (E) up to $t = 0.5$. Take

$$\Gamma = 4 \times 10^{-3} \mu m,$$

$$\mu = 10^{-3},$$

$$c^* = 7.52 \times 10^{-7} pg/(\mu m)^3,$$

$$c_0 = 1.05 c^*,$$

$$x^* = x(0) = .05 \mu m,$$

$$k = 5 \times 10^7,$$

so that $k\mu = 5 \times 10^4$ and $kc^* = 37.6$.

In this case kc_1 should turn out to be 45.73, and $\xi_1 = 2.16736 \times 10^{-2}$, $\xi_2 = 4.53479 \times 10^{-2}$. (Check this out.) Thus $x^* > \xi_2 > \xi_1$.

PROBLEM 1.10.3. Solve (E) up to $t = 0.5$. Take $x^* = .0975$ and all the other constants as in Problem 1.10.2.

In this case kc_1 should turn out to be 85.8229 while the roots ξ_1, ξ_2 should be

$$\xi_1 = 4.84721 \times 10^{-3},$$

$$\xi_2 = 9.771616 \times 10^{-2}.$$

Thus $\xi_1 < x^* < \xi_2$ here. Check this out.

PROBLEM 1.10.4. Solve (E) until $t = 0.16$. Take Γ, k, c^* as before but $x^* = .08$ and $k\mu = 5 \times 10^{-2}$. In this case kc_1 should turn out to be 39.736, while $\xi_1 = 8.3510 \times 10^{-2}$ and $\xi_2 = 1.197 \times 10^{-1}$. Note that $x^* < \xi_1 < \xi_2$.

The graphs for the solutions of Problems 1.10.2–1.10.4 are given in Fig. 1.4.

1.11. The Runge–Kutta method (BASIC)

While the Euler method could easily be motivated, it very often does not give the accuracy needed for scientific investigation. It is for that reason that many other methods are introduced—to give greater accuracy. A whole series of related methods going by the name of "Runge–Kutta" (dating from about the turn of the century) have the advantage that they are quite accurate and, at the same time, not too difficult to carry out. We are going to describe the most popular of these, leaving the motivation for later.

Again we look at the initial value problem

$$\frac{dy}{dt} = f(t, y), \qquad y(t_0) = y_0.$$

We will prescribe the approximate values of $y(t)$ by giving them only at the points t_k.

Suppose we know $y(t_k)$ and we want to determine an approximation y_{k+1} to $y(t_k + h)$. The idea of this method is to compute the value of $f(x, y)$ at several clearly chosen points near the solution curve in the interval $(t_k, t_k + h)$ and to combine these values so as to get a good value for $y_{k+1} - y_k$. We set

$$y_{k+1} = y_k + \tfrac{1}{6}\left(K_1 + 2K_2 + 2K_3 + K_4\right),$$

where

$$K_1 = hf(t_k, y_k),$$

$$K_2 = hf\left(t_k + \tfrac{1}{2}h,\ y_k + \tfrac{1}{2}K_1\right),$$

$$K_3 = hf\left(t_k + \tfrac{1}{2}h,\ y_k + \tfrac{1}{2}K_2\right),$$

$$K_4 = hf(t_k + h,\ y_k + K_3).$$

The simplest choice for t_k is given by $t_k = t_0 + kh$, while t_0 is the point at which the initial condition is given: $y(t_0) = y_0$.

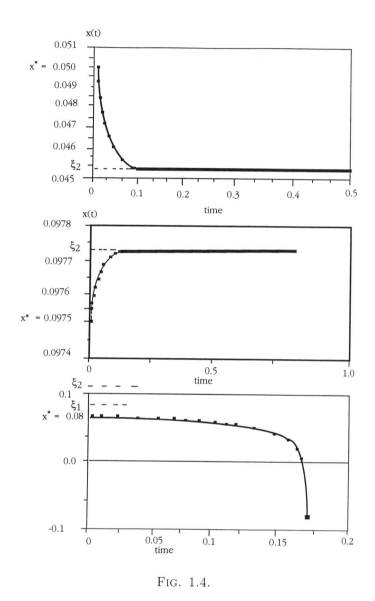

FIG. 1.4.

The case of a system of first-order differential equations can again be handled, as in the case of the Euler method, by writing the system in vector form

$$\frac{d\mathbf{y}}{dt} = f(t, \mathbf{y}), \qquad \mathbf{y}(t_0) = \mathbf{y}_0$$

and proceeding as in the case of a single equation. There is a slight notational difficulty if we want to represent the vectors $\mathbf{y}$ and $\mathbf{f}(t, \mathbf{y})$ by components, since subscripts are already used for the values of the approximate solution at different points. We use superscripts for components, such as

$$\mathbf{y}_k = \begin{pmatrix} y_k^{(1)} \\ y_k^{(2)} \\ \dots \\ y_k^{(m)} \end{pmatrix},$$

or, in more compact matrix notation, $\mathbf{y}_k = (y_k^{(1)}, \dots y_k^{(m)})^T$, where T represents the transpose of a row vector, which is a column vector. Here m represents the number of unknown functions, as well as the number of equations.

1.12. Discussion and motivation (BASIC)

In contrast to the Picard method of successive approximation, Euler's method and the Runge–Kutta method do not require repeated explicit integrations. On the contrary, if we take $f(t, y)$ to be *independent of y* these methods yield, as useful side products, methods of numerical integration, since the solution to the initial value problem

(1.20) $y'(t) = f(t), \qquad y(t_0) = y_0$

is nothing else than the definite integral

$$y(t) = y_0 + \int_{t_0}^{t} f(\tau)d\tau.$$

If we divide the t axis into intervals of size h, starting at $t = t_0$ and approximate $f(t)$ by the "step function" shown in Fig. 1.5, the resulting integration in the above formula will give us a piecewise linear continuous function $y_h(t)$ with $y_h(t_0) = y_0$, that is, *a polygonal curve*. The sum of the areas of the approximately "triangular" shaped pieces (between t_0 and t) represent the error $y(t) - y_h(t)$. It can be proved (and is intuitively clear) that this error approaches zero like the *first* power of h as $h \to 0$, or in the traditional terminology of the subject the error is "$O(h)$."

In the above discussion we approximated $f(t)$ by a piecewise constant function, and obtained an approximation $y_h(t)$ (for the solution $y(t)$ of problem (1.20)); $y_h(t)$ is a polygonal curve.

We can next try to approximate $f(t)$ by a polygon; see Fig. 1.6. The resulting integral is then the same as that given by what is called the trapezoidal rule. It seems plausible, on inspecting the diagram, that the error is much smaller as $h \to 0$, indeed it is $O(h^2)$.

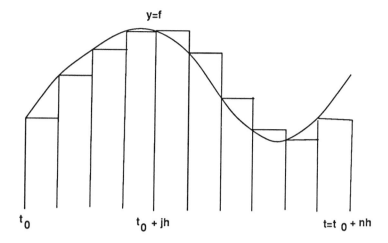

$$\textsc{Fig. 1.5.}$$

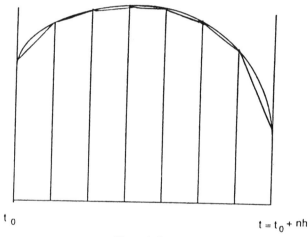

$$\textsc{Fig. 1.6.}$$

Now Simpson's rule is based on approximating $f(t)$ by a parabola in each interval of length $2h$. The basic identity here is

$$\int_{t_0}^{t_0+2h} f(t)dt = \frac{h}{3}\left[f(t_0) + 4f(t_0+h) + f(t_0+2h)\right] - \frac{h^5}{90} f^{(4)}(\xi),$$

where $t_0 < \xi < t_0 + 2h$.

By repeatedly applying this formula to neighboring intervals of length $2h$

we get "Simpson's rule" for evaluating integrals:

$$\int_{t_0}^{t_0+2nh} f(t)dt = \frac{h}{3}\left[f(t_0) + 4f(t_0 + h) + 2f(t_0 + 2h) + 4f(t_0 + 3h) + \cdots \right.$$

$$\left. +2f(t_0 + (2n - 2)h) + 4f(t_0 + (2n - 1)h) + f(t_0 + 2nh)\right] + O(h^4) \ .$$

This rule is valid provided f has four continuous derivatives. For cubic polynomials it holds exactly, without error term.

PROBLEM 1.12.1. Show that in the case where $f(t, y)$ is independent of y, the Runge–Kutta method as described above is equivalent to Simpson's rule with h replaced by $\frac{h}{2}$.

Note. It can be shown that the Runge–Kutta method also has an error of magnitude $O(h^4)$.

PROBLEM 1.12.2. Solve the problems in §1.7 by the Runge–Kutta method and compare with the results you obtained by the Euler method.

1.13. Crystals of several sizes

We consider the general case of crystals with N sizes. We assume that the initial concentration is larger than the critical concentration c^*

$$(1.21) \qquad\qquad c_0 > c^*.$$

We can then verify that the inequality

$$(1.22) \qquad\qquad c(t) > c^*$$

continues to hold for all $t > 0$. Indeed, if $c(t)$ becomes equal to c^* at some time t_0 then $dc/dt(t_0) \leq 0$. On the other hand,

$$(1.23) \quad \frac{dc}{dt} = -3\sum_j \mu_j x_j^2 \frac{dx_j}{dt} = -3\sum_j \mu_j x_j^2 G(x_j, c(t)) > 0 \qquad \text{(by (1.11))}$$

at $t = t_0$, a contradiction.

We also note that by (1.6), (1.9),

$$(1.24) \qquad\qquad c(t) < c_1$$

for all t, and therefore

$$(1.25) \qquad\qquad x_N(t) \leq \left(\frac{c_1}{\mu_N}\right)^{1/3}.$$

As in the case of a single-size crystal, some of the crystals may dissolve entirely in finite time, and they disappear thereafter from the differential equations.

Note that the size ordering of the solutions $x_j(t)$ continues to hold as long as the $x_j(t)$ are positive, that is, if $x_{j+1} - x_j > 0$ and $x_j > 0$ at time t, then the inequality $x_{j+1} - x_j > 0$ continues to hold as long as x_j remains positive. Indeed, this follows from

$$\frac{d}{dt}(x_{j+1} - x_j) = G(x_{j+1}, c(t)) - G(x_j, c(t)) > 0.$$

The last inequality actually implies that

(1.26) $x_{j+1}(t) - x_j(t)$ is strictly monotone increasing.

The curve $x = L^*(t)$ determines whether a crystal grows or shrinks. If $x_j(t) > L^*(t)$ then $x_j(t)$ is growing, whereas if $x_j(t) < L^*(t)$ then $x_j(t)$ is shrinking.

We denote by k the maximal number of crystal sizes that have disappeared in finite time, say by the time $t = t_0$. Thus for $t > t_0$ there are only crystal grains with sizes

(1.27) $x_{k+1}(t), \ldots, x_N(t)$

present in the differential systems (the $x_1, \ldots, x_k$ dropped out).

THEOREM 1.13.1. *All crystals that do not have the largest size x_N will dissolve in finite time, i.e., $k + 1$ cannot be smaller than N (in (1.27)).*

Proof. We suppose that $k + 1 < N$ and derive a contradiction. A crystal of size $x_j(t)$ may possibly intersect the curve $x = L^*(t)$ several times. However, $x_N(t)$ can intersect $x = L^*(t)$ at most once. Indeed, at a point of intersection $t = t_*$ we have

$$\frac{dx_N}{dt} = G(x_N, c(t)) = 0$$

whereas

$$\frac{dL^*}{dt} < 0 \quad \text{since} \quad \frac{dc}{dt} > 0$$

as seen from (1.23) (the sum ranges over $k + 1 \leq j \leq N$, and each G with $k + 1 \leq j < N$ is negative).

We conclude that for some $t_* > t_0$ either

(1.28) $x_N(t) > L^*(t)$ for all $t > t_*$

or

(1.29) $x_N(t) < L^*(t)$ for all $t > t_*$.

If (1.29) holds then

$$\frac{dx_N}{dt} < 0 \quad \text{for} \quad t > t_*,$$

whereas if (1.28) holds then

$$\frac{dx_N}{dt} > 0 \quad \text{for all } t > t_*.$$

In both cases $\lim x_N(t)$ exists (recall that $x_N(t)$ is bounded, by (1.25)).

From (1.26) with $j + 1 = N$ it follows that $\lim_{t \to \infty} x_{N-1}(t)$ also exists, and similarly $\lim_{t \to \infty} x_j(t)$ exists for all $N \geq j \geq k + 1$.

From the definition of $c(t)$ in (1.6) it also follows that $\lim_{t \to \infty} c(t)$ exists. Set

$$x_j(\infty) = \lim_{t \to \infty} x_j(t) , \qquad c(\infty) = \lim_{t \to \infty} c(t).$$

By (1.26),

(1.30) $$x_{k+1}(\infty) < x_N(\infty).$$

Hence

$$G(x_{k+1}(\infty) , c(\infty)) < G(x_N(\infty), c(\infty)).$$

It follows that at least one of these two numbers is different from zero; say

$$\eta \equiv G(x_{k+1}(\infty), c(\infty)) \neq 0.$$

Since for large t

$$\frac{dx_{k+1}(t)}{dt} = G(x_{k+1}(t), c(t)) \sim \eta,$$

we deduce that $\lim_{t \to \infty} x_{k+1}(t)$ does not exist, which is a contradiction. The contradiction can be avoided only if $k + 1 = N$, and thus Theorem 1.13.1 follows.

We now know that

$$x_1(t) \quad \text{dissolves in time } \tau_1,$$

$$x_2(t) \quad \text{dissolves in time } \tau_2,$$

$$\cdots$$

$$x_{N-1}(t) \quad \text{dissolves in time } \tau_{N-1}.$$

For $t > \tau_{N-1}$ we are back into the one-size crystals situation studied in §1.8.

PROBLEM 1.13.1. Compute $\tau_1, \tau_2, \ldots, \tau_{N-1}$. (Take $N = 2$ and use the physical constants suggested earlier.)

PROBLEM 1.13.2. Compute $c(\infty)$. Is $c^* < c(\infty) < c_1$? The answer may depend on the initial conditions; investigate.

1.14. Summary

We have analyzed a model of crystal growth in a solution by theoretical methods in ordinary differential equations, actually proving some mathematical theorems on the behavior of crystals in a solution. We have also verified these theoretical results by numerical calculations and graphed the results. We have concluded that after a finite time $t = \tau$ all but the largest size crystals will have dissolved in the solution. Thereafter, the remaining crystals, having uniform size, will either (i) shrink to zero size (and hence dissolve) in finite time; or (ii) converge to one of two sizes ξ_1 or ξ_2, depending on the conditions existing at time $t = \tau$. Numerical methods for solving ordinary differential equations allow us to compute the limiting size of the crystals.

References

The Ostwald ripening model is described in

[1] N. S. Tavare, *Simulation of Ostwald ripening in a reactive batch crystallizer*, AiChEJ, 33 (1987), pp. 152–156.

[2] A. Friedman, *Mathematics in Industrial Problems*, Chap. 4, IMA Vol. Math. Appl., Vol. 16, Springer–Verlag, New York, 1988.

The analysis in §1.13 is taken from

[3] A. Friedman, B. Ou and D. Ross, *Crystal precipitation with discrete initial data*, J. Math. Anal. Appl., 137 (1989), pp. 576–590.

Chapter 2

Air Quality Modeling

2.1. Background

Air quality has become an important societal issue. Acid rain is a regional problem, affected by industrial by-products of toxic gas; it pollutes the ground and damages vegetation. In urban areas it is the ozone concentration that is considered to be the biggest health hazard. Air quality models are mathematical descriptions of atmospheric transport, diffusion, and chemical reaction of pollutants. The unknown variables are concentrations of chemical species in the air. The aim in developing and studying such models is to be able to predict how peak concentrations will change in response to prescribed changes in meteorology and in the source of pollution. Ozone air quality modeling has been one of the main areas of emphasis in the United States in the last twenty years; it is of particular interest to the automobile industry. In this chapter we consider the modeling of transport and diffusion of a single chemical, say ozone, ignoring the various underlying processes.

Putting the problem in a more personal context, suppose there is an industrial plant emitting noxious fumes. Also suppose that once an hour, as part of the manufacturing process, the plant emits a very concentrated batch of these offensive fumes for a few minutes and then stops. Exactly one mile away there is a beautiful house you would like to buy. If the wind is blowing in the direction of the house (the "worst possible" direction), how objectionable will the fumes be by the time they reach "your" house? In other words, what will be the maximum concentration of the fumes as they pass your house?

Assuming there are no chemical changes taking place as the fumes travel, there are basically two processes at work here: *advection* and *diffusion.*

Advection is essentially the effect of the wind "blowing" the fumes in a given direction without significantly dispersing them. A good example is a distant cloud moving with a fixed velocity in a given direction without apparently altering its size or shape.

Let us first consider the one-dimensional situation where there is advection but no diffusion (see Fig. 2.1).

FIG. 2.1.

Suppose at time $t = 0$ the density of the fumes has a distribution given by profile $c_0(x)$ shown in Fig. 2.2.

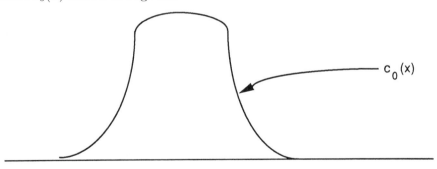

FIG. 2.2.

This profile moves to the right with the constant wind velocity U, giving rise to the moving profile for the concentration

$$(2.1) \qquad\qquad c(x,t) = c_0(x - Ut).$$

Differentiating partially, using the chain rule, we get

$$\frac{\partial c}{\partial x}(x,t) = c_0'(x - Ut), \qquad \frac{\partial c}{\partial t}(x,t) = c_0'(x - Ut) \cdot (-U),$$

thus giving us the "advection equation"

$$(2.2) \qquad \frac{\partial c}{\partial t}(x,t) + \frac{\partial (Uc)}{\partial x}(x,t) = 0$$

with "initial" condition $c(x,0) = c_0(x)$. In the particular situation described here, we knew the solution of the partial differential equation in advance. In a more complicated situation, such as when the "wind velocity" U is not a constant, this will not be the case.

Looking at the form with the concentration profile (2.1) we see that the noxious fumes will arrive at "your" house in the same concentration as they left the factory: very high! This is bad news indeed.

Luckily all is not lost! As we have hinted earlier, there is another process at work: *diffusion*. This is the reason that even without the presence of wind, foul-smelling odors usually disappear after a time. For example, a newly painted house stops smelling of paint after a day or two, if not earlier. Taking into account advection and diffusion, there is at least the possibility that by the time the fumes have reached your house, diffusion has had a big enough effect, so that the fumes are barely noticeable.

In the remainder of this chapter we will give a more detailed mathematical description of the processes of advection and diffusion, how these can be investigated numerically, and finally how the two processes combined can be described by a simple "advection diffusion equation" that can be studied numerically.

2.2. The model

We denote by c the concentration of one species; it is a function of position (x_1, x_2, x_3) and of time t. The species is being transported by the wind, whose velocity $\vec{u} = \vec{u}(x_1, x_2, x_3, t)$ is assumed to be known. Particles of the species are also diffusing locally; they tend to move from areas of high concentration to areas of low concentration. If diffusion is ignored then the transport equation is

$$(2.3) \qquad \frac{\partial c}{\partial t} + \nabla \cdot (\vec{u}\, c) = 0.$$

This is in some contexts also called the *continuity equation*. If we integrate (2.3) over any bounded domain D in $\mathbb{R}^3$ we get

$$(2.4) \qquad \frac{d}{dt} \iiint_D c(x_1, x_2, x_3, t) dx_1 dx_2 dx_3 = - \iint_{\partial D} c\, \vec{u} \cdot \vec{n}\; dS,$$

where ∂D is the boundary of D and $\vec{n}$ is the outward unit normal to ∂D. This equation says that the rate of increase of the chemical in any domain D is equal to the flow of chemicals across the boundary. Conversely, if (2.4) holds for any domain D then, upon taking a sequence of domains D_j shrinking to a point, we can recover the relation (2.3) at that point.

If diffusion is not ignored then (2.3) is replaced by a more complicated partial differential equation:

$$(2.5) \qquad \frac{\partial c}{\partial t} + \nabla \cdot (\vec{u}\, c) = \sum_{i,j=1}^{3} \frac{\partial}{\partial x_i}\left(k_{ij}\frac{\partial c}{\partial x_j}\right),$$

where (k_{ij}) is a positive definite matrix, called the *diffusion matrix*.

In either case, (2.3) or (2.5), we are given the concentration c at an initial time, say at $t = 0$,

$$(2.6) \qquad c(x_1, x_2, x_3, 0) = c_0(x_1, x_2, x_3)$$

and our task is to compute the concentration $c(x_1, x_2, x_3, t)$ at subsequent times. In particular we wish to find out the maximum values of the concentration at a prescribed time; government regulations for pollution control often take the maximum concentration of pollution as a critical factor.

2.3. The advection equation

In this section we ignore diffusion and assume that the wind velocity is only in the horizontal direction. For simplicity we also assume that the direction of the wind is fixed, say, in the x-direction. Then $\vec{u} = (U, 0, 0)$ and the transport equation reduces to

$$(2.7) \qquad \frac{\partial c}{\partial t} + \frac{\partial(Uc)}{\partial x} = 0;$$

this is called the *advection equation*. We also assume that initially c depends only on x, i.e.,

$$(2.8) \qquad c(x, 0) = c_0(x), \qquad -\infty < x < \infty.$$

The velocity $U = U(x)$ is a function of x.

To solve (2.7), (2.8), we rewrite (2.7) in the form

$$(2.9) \qquad \frac{\partial c}{\partial t} + U\frac{\partial c}{\partial x} = f \qquad (f = -U_x c)$$

and assume that $U(x)$ is continuously differentiable ($U_x = dU/dx$).

Consider the differential equation

$$(2.10) \qquad \begin{cases} \dfrac{dx}{dt} = U(x), & t > 0, \\[2mm] x(0) = x_0 \end{cases}$$

and denote its solution $x(t)$ by $x(t; x_0)$. Geometrically, $x(t; x_0)$ determines a unique curve γ_{x_0} passing through the point $(x_0, 0)$. We can show that $x(t; x_0)$ is actually differentiable in the parameter x_0 and the derivative

$$z(t) \equiv \frac{\partial x(t; x_0)}{\partial x_0}$$

satisfies

$$\frac{dz}{dt} = U_x(x(t; x_0))z, \qquad z(0) = 1.$$

PROBLEM 2.3.1. Prove the last statement.

We now examine the function

$$c(x(t; x_0), t)$$

as a function of the variable t. We find that

$$\frac{dc}{dt} = \frac{\partial c}{\partial t} + \frac{\partial c}{\partial x}\frac{dx}{dt} = \frac{\partial c}{\partial t} + U\frac{\partial c}{\partial x} = f = -U_x(x(t; x_0))c,$$

or

$$\frac{d}{dt}\log c = -U_x(x(t; x_0)).$$

It follows that

$$(2.11) \qquad c(x(t; x_0), t) = c_0(x_0)\exp\left\{-\int_0^t U_x(x(s; x_0))ds\right\}.$$

Note that d/dt is simply differentiation along the curve γ_{x_0}, parametrized by t.

We have shown that the solution of (2.7), (2.8) must be given by the formula (2.11). Conversely we can show that if the curves $(x(t; x_0), t)$ cover the upper half $(t \geq 0)$ of the (x, t) plane, then the right-hand side of (2.11) is a solution to (2.7), (2.8).

PROBLEM 2.3.2. Prove the last statement.

DEFINITION 2.3.1. The curves (2.10) are called *characteristics* of (2.9) (for any f). The method described above for computing the solution c is called the *method of characteristics*.

2.4. Numerical methods

Although (2.11) provides a very nice formula for the solution, it is not very useful for air modeling computations. Indeed, in real-life situation we are given a finite number of locations of air quality test stations, say at $X_1, X_2, \ldots, X_N$. We are assigned the task of evaluating c at each location X_j at certain times $T_1, T_2, \ldots$. To compute c at (X_j, T_1), say, (2.11) will require us to discover the point x_0 such that the characteristic through $(x_0, 0)$ passes through (X_j, T_1). If we are to compute c at a later time T_2, we need to discover the corresponding (new) point x_0. For large times T_N, the computation of the corresponding point x_0 may be quite time consuming.

Therefore, we must develop a better computational method. It is based on *finite differences*. We divide the x-space into intervals of equal length Δx and the positive time axis into intervals of equal length Δt. We wish to approximate the values $c(j\Delta x, n\Delta t)$ by some quantities c_j^n satisfying certain approximate equations.

For simplicity we first consider the case when U is independent of x, so that (2.7) and (2.8) become

$$(2.12) \qquad \frac{\partial c}{\partial t} + U\frac{\partial c}{\partial x} = 0, \qquad c(x, 0) = c_0(x).$$

To facilitate solving the "initial value problem" (2.12) numerically, we replace derivatives by finite differences. However, before we can do that we have to introduce a set of "lattice points" in the $x - t$ plane given by $x = j\Delta x$, $t = n\Delta t$, $j = 0, \pm 1, \pm 2, \ldots$, $n = 0, 1, 2, 3 \ldots$. The approximation to $c(j\Delta x, n\Delta t)$ will be denoted by c_j^n, as already mentioned above. Note that the superscript "n" is not a power. We approximate

$$\frac{\partial c}{\partial t} \quad \text{by} \quad \frac{c(j\Delta x, (n+1)\Delta t) - c(j\Delta x, n\Delta t)}{\Delta t}$$

or rather, $(c_j^{n+1} - c_j^n)/\Delta t$, and

$$\frac{\partial c}{\partial x} \quad \text{by} \quad \frac{c(j\Delta x, n\Delta t) - c((j-1)\Delta x, n\Delta t)}{\Delta x}$$

or rather, $(c_j^n - c_{j-1}^n)/\Delta x$. Then (2.12) is replaced by

$$\frac{c_j^{n+1} - c_j^n}{\Delta t} + U\frac{c_j^n - c_{j-1}^n}{\Delta x} = 0$$

or

$$(2.13) \qquad c_j^{n+1} = c_j^n - \frac{U\Delta t}{\Delta x}(c_j^n - c_{j-1}^n).$$

Since c_j at "time" $n+1$ is given explicitly by the c_j at time n, we refer to (2.13) as an *explicit* finite difference scheme. The finite differences scheme described above is called "forward in t, backward in x." (Can you guess why?)

From (2.13) it is clear that if the values of the c's are known in row $n = 0$ then they are *explicitly* known in row $n = 1$. Repeating this process, we can then determine all the c_j^n for the initial value c_j^0. See Fig. 2.3.

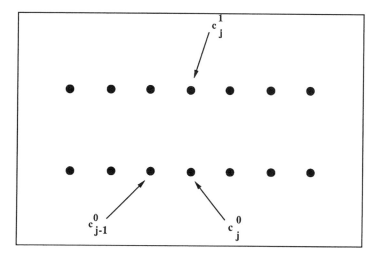

FIG. 2.3.

Two questions arise: First, if Δx and Δt are made sufficiently small, do the c_j^n's approximate values $c(j\Delta x,\ n\Delta t)$ at the mesh points in some sense? Second, keeping Δx and Δt fixed, do the c_j^n's remain bounded as $n \to \infty$ uniformly in the j's (or in some other sense)? If the answer to the first question is in the affirmative we say the finite difference scheme is *convergent*. If the answer to the second question is affirmative we call the scheme *stable*. Note that the second question really concerns the behavior of the solution to the discretized problem as the (discretized) time variable becomes very large: Does the solution stay bounded, or does it blow up as $n \to \infty$?

It would seem that the convergence question is the more relevant one. However, it turns out that stability is easier to check. Luckily there is a theorem that says, roughly speaking, that *if the original* (differential) *initial value problem is well-posed* (for example if it does a good job of representing the physical situation), *and if one has done a reasonable job of replacing the derivatives by finite differences, then stability implies convergence.*

The quantity

(2.14)
$$\sigma = \frac{U\Delta t}{\Delta x}$$

plays a fundamental role in the scheme (2.13). It turns out that if $0 < \sigma \le 1$

then the numerical scheme based on (2.13) converges. However, if $\sigma > 1$ then the scheme does not converge, i.e., as Δx and Δt become smaller the finite difference solution develops larger oscillations. The scheme is *stable*, if $0 < \sigma \leq 1$ and *unstable* if $\sigma > 1$; see §2.8.

PROBLEM 2.4.1. Let

$$c_0(x) = \begin{cases} 5 & \text{if} \quad |x| < 1 \\ 0 & \text{if} \quad |x| > 1. \end{cases}$$

Compute c by the scheme (2.13) in the cases where $U = 1, 5, 10, 20, 40$. Find the maximum concentration at time $t = 4$. How does it depend on U?

PROBLEM 2.4.2. Let

$$c_0(x) = \begin{cases} \sin^2 \pi(x+2) & \text{if} \quad -2 < x < -1 \quad \text{or} \quad 1 \leq x \leq 2 \\ 0 & \text{otherwise} \end{cases}$$

and take $U = 1$. Compute the profile of $c(x, t)$ for times $t = 1, 2, 3, \ldots, 10$.

PROBLEM 2.4.3. Do the same for

$$c_0(x) = \begin{cases} \sin^2 x & \text{if } 0 < x < \pi \\ 0 & \text{otherwise.} \end{cases}$$

The explicit finite difference scheme can be extended to (2.7) with general $U(x)$; it has the form

$$(2.15) \qquad c_j^{n+1} = c_j^n - \frac{U_j \Delta t}{\Delta x}(c_j^n - c_{j-1}^n) - \frac{\Delta t}{\Delta x}(U_j - U_{j-1})c_j^n,$$

where $U_j = U(j\Delta x)$.

PROBLEM 2.4.4. Use the scheme (2.15) to solve (2.7), (2.8) for c_0 as in Problem 2.4.1 when

$$U(x) = \begin{cases} 0 & \text{if} \quad x \leq 0 \\ \dfrac{x^2}{1+x^2} & \text{otherwise.} \end{cases}$$

What is the profile of c at times $t = 1, 2, 4, 10$?

Note 1. The concentration of a chemical is always a nonnegative quantity. This is very nicely seen from (2.11): if initially $c_0 \geq 0$ then $c \geq 0$ for all future times. However, this result may not necessarily hold at the approximation level of (2.13). That is, even if $c_0 \geq 0$, it may happen that c_i^n becomes negative for some indices n, i. Of course, as Δt and Δx decrease to zero, the values of the negative quantities c_i^n must tend to zero when the scheme is convergent.

2.5. The general advection equation

So far we have assumed that the wind velocity was along the horizontal x-direction. We now consider the more realistic situation where the wind velocity is in any horizontal direction. This means that

$$\vec{u} = (U, V, 0).$$

The advection equation is then

$$(2.16) \qquad \frac{\partial c}{\partial t} + \frac{\partial}{\partial x}(Uc) + \frac{\partial(Vc)}{\partial y} = 0$$

or

$$(2.17) \qquad \frac{\partial c}{\partial t} + U\frac{\partial c}{\partial x} + V\frac{\partial c}{\partial y} + c\frac{\partial U}{\partial x} + c\frac{\partial V}{\partial y} = 0.$$

We take initial conditions (independent of z)

$$(2.18) \qquad c(x, y, 0) = c_0(x, y).$$

PROBLEM 2.5.1. Generalize the method of characteristics.

The explicit finite difference scheme (2.13) generalizes to

$$(2.19) \qquad c_{j,\ell}^{n+1} = c_{j,\ell}^n - \frac{U\Delta t}{\Delta x}\left(c_{j,\ell}^n - c_{j-1,\ell}^n\right) - \frac{V\Delta t}{\Delta y}\left(c_{j,\ell}^n - c_{j,\ell-1}^n\right).$$

Equation (2.15) can be similarly extended.

Consider the initial concentration in $\mathbb{R}^2$

$$c_0(x, y) = \begin{cases} 50\left(1 + \cos\dfrac{\pi R}{4}\right) & \text{if } R < 4 \\ 0 & \text{if } R > 4, \end{cases}$$

where

$$R^2 = (x - x_0)^2 + (y - y_0)^2, \qquad (x_0, y_0) = (5, -10).$$

The concentration is referred to as a cosine hill.

We take a wind velocity to be uniform intensity 1, rotating counterclockwise,

$$(U, V) = (-\sin\theta, \cos\theta),$$

where $\theta = \arctan y/x$.

PROBLEM 2.5.2. Find the profile of c for $|x| \leq 100$, $|y| \leq 100$, $t = 3$. (See Figs. 2.4 and 2.5.)

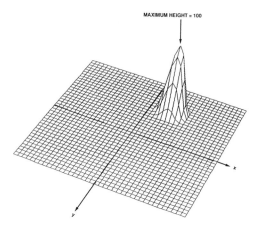

FIG. 2.4. *Three-dimensional graph of the solution to Problem 2.5.2 ($t = 0$).*

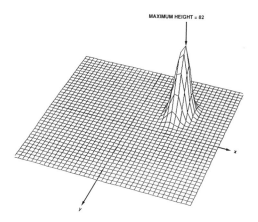

FIG. 2.5. *Three-dimensional graph of the solution to Problem 2.5.2 ($t = 5$).*

2.6. Enters diffusion

We now add diffusion to the advection equation (2.7):

$$(2.20) \qquad \frac{\partial c}{\partial t} + \frac{\partial}{\partial x}\,(Uc) = k\frac{\partial^2 c}{\partial x^2} \qquad (k > 0)$$

and again try to solve (2.20) with an initial condition

$$(2.21) \qquad c(x,0) = c_0(x) \qquad (-\infty < x < \infty).$$

The method of characteristics is no longer valid. However, the finite

difference scheme is still applicable.   Motivated by Taylor's formula, we
approximate

$$\frac{\partial^2 c}{\partial x^2} \quad \text{at} \quad (j\Delta x, n\Delta t) \quad \text{by}$$

$$\frac{1}{(\Delta x)^2} \{c((j+1)\Delta x, n\Delta t) - 2c(j\Delta x, n\Delta t) + c((j-1)\Delta x, n\Delta t))\}$$

and $\partial c/\partial t$, $\partial c/\partial x$ as before. We get

$$(2.22) \quad c_j^{n+1} = c_j^n - \frac{U\Delta t}{\Delta x} (c_j^n - c_{j-1}^n) + k\frac{\Delta t}{(\Delta x)^2} (c_{j+1}^n - 2c_j^n + c_{j-1}^n).$$

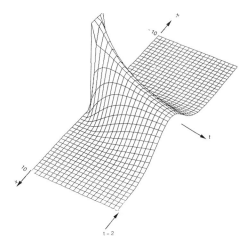

FIG. 2.6.   *Three-dimensional graph of the solution to Problem* 2.6.1. $U = 1, r = .25, \Delta x = .1, \Delta t = 2.5 \times 10^{-3}$ (*here* $r = \Delta t/(\Delta x)^2$.)

PROBLEM 2.6.1. Let $k = 1$ and consider the scheme (2.22) with the same c_0 and U as in Problem 2.4.1. Compute the maximum concentration at time $t = 4$ and compare with the results obtained in Problem 2.4.1. (See Figs. 2.6 and 2.7.) Does the diffusion tend to decrease the maximum concentration?

PROBLEM 2.6.2. Do the same for Problem 2.4.4.

2.7. The von Neumann stability criterion

In §2.4 we defined the concepts of convergence and stability for finite difference schemes; in §2.9 we define the concept of *consistent* scheme. It is known that a difference scheme is convergent if it is both stable and consistent. Finite difference schemes that are motivated by Taylor's formula are usually consistent. Stability, however, is not always satisfied and is often hard to check

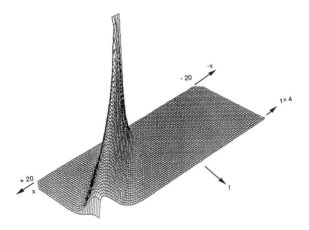

FIG. 2.7.   *Three-dimensional graph of the solution to Problem* 2.6.1. $U = 5, r = .25, \Delta x = .1, \ \Delta t = 2.5 \times 10^{-3}$.

directly. In this section we introduce the von Neumann criterion for stability and then apply it to some examples; this condition is relatively easy to check since it assumes a special form of the solution of the finite difference scheme.

This criterion states that *the difference method for an initial value problem (for a differential equation with constant coefficients) with a bounded solution is stable if every solution to the finite difference equation having the form*

$$c_j^n = \xi^n e^{i\beta j}$$

$$(\beta \ \ real, \quad \xi = \xi(\beta) \ \ complex)$$

has the property $|\xi| \le 1$. *Note that* ξ^n *here means* ξ *raised to the* nth *power!*

Let us apply the von Neumann criterion to (2.22).

Letting $c_j^n = \xi^n e^{i\beta j}$, $r = \Delta t/(\Delta x)^2$ in (2.22) gives us

$$\xi^{n+1} e^{i\beta j} = \xi^n e^{i\beta j} - Ur \cdot \Delta x (\xi^n e^{i\beta j} - \xi^n e^{i\beta(j-1)})$$
$$+ kr(\xi^n e^{i\beta(j+1)} - 2\xi^n e^{i\beta j} + \xi^n e^{i\beta(j-1)}).$$

Hence,

$$\xi = 1 - U \cdot r \cdot \Delta x (1 - e^{-i\beta}) + k \cdot r \cdot (e^{i\beta} - 2 + e^{-i\beta})$$

$$= 1 - U \cdot r \cdot \Delta x (1 - \cos \beta) - iUr\Delta x \sin \beta + 2kr(\cos \beta - 1)$$

$$= \left(1 - (2kr + Ur\Delta x)(1 - \cos \beta)\right) - iUr\Delta x \sin \beta.$$

Thus

$$|\xi|^2 = 1-2(2kr+Ur\Delta x)(1-\cos\beta)+(2kr+Ur\Delta x)^2(1-\cos\beta)^2+U^2r^2(\Delta x)^2\sin^2\beta,$$

$$(2.23) \quad |\xi|^2 - 1 = -2r(2k + U\Delta x)(1 - \cos\beta) + 4k^2r^2(1 - \cos\beta)^2$$
$$+ 4kUr^2\Delta x(1 - \cos\beta)^2 + U^2r^2(\Delta x)^2(1 - 2\cos\beta + 1).$$

Now suppose for the moment that $(1 - \cos\beta) > 0$. Then the condition $|\xi|^2 \le 1$ is equivalent to

$$-(2k + U\Delta x) + 2k^2r(1 - \cos\beta) + 2kUr\Delta x(1 - \cos\beta) + U^2r(\Delta x)^2 \le 0,$$

or to

$$r\left(U^2(\Delta x)^2 + (2k^2 + 2kU\Delta x)(1 - \cos\beta)\right) \le 2k + U\Delta x.$$

The condition that this hold for *all real* β (with $1 - \cos\beta > 0$) is equivalent to

$$r(U^2(\Delta x)^2 + 4kU\Delta x + 4k^2) = r(U\Delta x + 2k)^2 \le 2k + U\Delta x,$$

and hence to
$$(2.24) \qquad\qquad\qquad r(2k + U\Delta x) \le 1.$$

These inequalities show that if the von Neumann criterion holds, i.e., if $|\xi| \le 1$ for *all* real β (with $1 - \cos\beta > 0$), then (2.24) holds. Conversely if (2.24) holds under the condition that $1 - \cos\beta > 0$, then $|\xi|^2 \le 1$. However, the only way that $1 - \cos\beta = 0$ can hold in (2.23) is if $|\xi| = 1$, which would still satisfy the von Neumann criterion.

Let us apply the von Neumann criterion to solve Problem 2.6.1. Taking $\Delta x = 0.1$, $k = 1$, and $U = 1$, we get the stability criterion satisfied for $\Delta t \le 1/210$. The results of the computations for $t = 4$ are shown in Figs. 2.8, 2.9, and 2.10 for values $\Delta t = 1/211$ (stable), $\Delta t = 1/210$ ("barely stable") and $\Delta t = 1/209$ (unstable).

We can make up two theories at this point.

Optimist's Theory. The above computations show how sensitive the von Neumann criterion is.

Pessimist's Theory. This was just a lucky coincidence.

PROBLEM 2.7.1. By doing more computations investigate which theory is more likely to be correct.

PROBLEM 2.7.2. For the equation $c_t + Uc_x = 0$ ($U = $ constant) investigate the stability of the explicit scheme (forward in t, forward in x)
$$\frac{c_j^{n+1} - c_j^n}{\Delta t} + U\frac{(c_{j+1}^n - c_j^n)}{\Delta x} = 0.$$

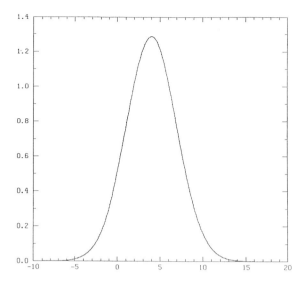

FIG. 2.8. *Solution to Problem 2.6.1 with $U = 1, \Delta x = 1/10, \Delta t = 1/211$ (stable).*

PROBLEM 2.7.3. Do the same for the explicit scheme (forward in t, backward in x)

$$\frac{c_j^{n+1} - c_j^n}{\Delta t} + \frac{U(c_j^n - c_{j-1}^n)}{\Delta x} = 0.$$

PROBLEM 2.7.4. Do the same for the explicit scheme

$$\frac{c_j^{n+1} - \frac{1}{2}(c_{j+1}^n + c_{j-1}^n)}{\Delta t} + U\frac{(c_{i+1}^n - c_{j-1}^n)}{2\Delta x} = 0.$$

PROBLEM 2.7.5. Investigate the stability of the Lax–Wendroff scheme

$$c_j^{n+1} = c_j^n - \frac{U}{2}\frac{\Delta t}{\Delta x}(c_{j+1}^n - c_{j-1}^n) + \frac{U^2}{2}\left(\frac{\Delta t}{\Delta x}\right)^2(c_{j+1}^n - 2c_j^n + c_{j-1}^n).$$

PROBLEM 2.7.6. Use the Lax–Wendroff scheme to attack Problem 2.4.1.

2.8. Stability, consistency, and convergence

One of the useful facts in studying difference methods for initial value problems (for a differential equation with constant coefficients) is that they can be studied by considering initial values of the form

(2.25) $$c_0(x) = e^{i\kappa x}$$

(for κ real) only. In this case, the exact solution is of the form

(2.26) $$c(x, t) = e^{vt}e^{i\kappa x}$$

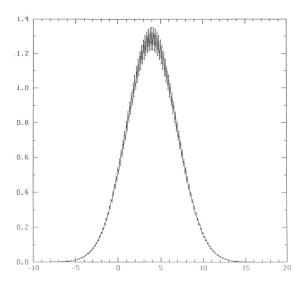

FIG. 2.9. *Solution to Problem* 2.6.1 *with* $U = 1, \Delta x = 1/10, \Delta t = 1/210$ (*"barely stable"*).

(for some complex constant v), and the solution given by the difference scheme is of the form

$$(2.27) \qquad c_j^n = \xi^n e^{i\beta j},$$

where $\beta = \kappa \Delta x$. (Note that, as in the previous section, ξ^n means "ξ raised to the nth power.") For example, inserting (2.26) in (2.20), we see that

$$v = -iU\kappa - k\kappa^2,$$

and hence, that the solution of (2.20) satisfying the initial condition (2.25) is

$$c(x, t) = e^{-(iU\kappa + \kappa^2 k)t} e^{i\kappa x}.$$

To obtain the solution of the scheme (2.22) satisfying the initial condition (2.25) at the grid points $x = j\Delta x$, we insert (2.27) into (2.22) and get (with $r = \Delta t/(\Delta x)^2$),

$$\xi^{n+1} e^{i\beta j} = \xi^n e^{i\beta j} - U r \Delta x \left(\xi^n e^{i\beta j} - \xi^n e^{i\beta(j-1)} \right)$$

$$+ k \left(\xi^n e^{i\beta(j+1)} - 2\xi^n e^{i\beta j} + \xi^n e^{i\beta(j-1)} \right).$$

Hence, as in the previous section,

$$\xi = 1 - (2kr + U r \Delta x)(1 - \cos \beta) - iU r \Delta x \sin \beta.$$

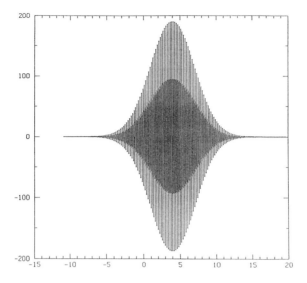

FIG. 2.10.   *Solution to Problem 2.6.1 with $U = 1, \Delta x = 1/10, \Delta t = 1/209$* (*unstable*).

In the general situation, the error made by using the solution given by the difference scheme instead of the exact solution at $(x, t) = (j\Delta x, n\Delta t)$ is

$$
\begin{aligned}
e_j^n &= c(j\Delta x, n\Delta t) - c_j^n \\
&= e^{v\Delta t \cdot n} e^{i\kappa x} - \xi^n e^{i\kappa x} \\
&= \left((e^{v\Delta t})^n - (\xi^n) \right) e^{i\kappa x}
\end{aligned}
$$

and so,

$$
|e_j^n| \le |(e^{v\Delta t})^n - \xi^n|.
$$

Since

$$
(e^{v\Delta t})^n - \xi^n = \left(e^{v\Delta t} - \xi \right)\left(e^{v\Delta t(n-1)} + e^{v\Delta t(n-2)}\xi + \cdots + e^{v\Delta t}\xi^{n-2} + \xi^{n-1} \right),
$$

we have

$$
|(e^{v\Delta t})^n - \xi^n| \le C_{v,n} C_{\xi,n} |e^{v\Delta t} - \xi| n,
$$

where

$$
C_{v,n} = \sup_{0 \le \ell \le n-1} |e^{(v\Delta t)\ell}|,
$$

$$
C_{\xi,n} = \sup_{0 \le \ell \le n-1} |\xi^\ell|.
$$

Since $n = t/\Delta t$, we have

$$|e_j^n| \leq C_{v,n} C_{\xi,n} t |\frac{e^{v\Delta t} - \xi}{\Delta t}|.$$

If the exact solution is bounded for all t, then we must have that $C_{v,n}$ is smaller than some fixed constant C_v for every $n \in \mathbb{N}$ ($\mathbb{N}$ = positive integers). We thus have

(2.28) $$|e_j^n| \leq C_v \, C_{\xi,n} t |\frac{e^{v\Delta t} - \xi}{\Delta t}|.$$

A numerical scheme satisfies the stability criterion if $C_{\xi,n}$ is uniformly bounded with respect to n, that is, if

$$|\xi^n| \leq C_\xi \quad \forall \ n \in \mathbb{N},$$

i.e., if $|\xi| \leq 1$, which is precisely the von Neumann criterion.

A numerical scheme is said to be *consistent* if as Δx and Δt go to zero, we have

$$|\frac{e^{v\Delta t} - \xi}{\Delta t}| \to 0.$$

A numerical scheme is said to be *convergent* if as Δx and Δt go to zero, $|e_j^n|$ also goes to zero. (When the scheme is convergent the finite difference solutions approximate the exact solution when Δx and Δt go to zero.)

The inequality (2.28) relates the notions of stability, consistency, and convergence: *A numerical scheme that is consistent is convergent if it is stable.*

Let us see if the difference scheme (2.22) is consistent. Since

$$e^{v\Delta t} = e^{-(iU\kappa + k\kappa^2)\Delta t}$$

$$= 1 - (iU\kappa + k\kappa^2)\Delta t + O((\Delta t)^2),$$

and

$$\xi = 1 - \left(2k\frac{\Delta t}{(\Delta x)^2} + U\frac{\Delta t}{\Delta x}\right)(1 - \cos(\kappa\Delta x)) - iU\frac{\Delta t}{\Delta x} \sin(\kappa\Delta x)$$

$$= 1 - \left(2k\frac{\Delta t}{(\Delta x)^2} + U\frac{\Delta t}{\Delta x}\right)\frac{\kappa^2}{2} \Delta x^2 \left(1 + O(\kappa^2(\Delta x)^2)\right)$$

$$-iU\frac{\Delta t}{\Delta x}(\kappa\Delta x)\left(1 + O(\kappa^2(\Delta x)^2)\right)$$

$$= 1 - (iU\kappa + k\kappa^2)\Delta t + \Delta t(O(\Delta x)),$$

we obtain

$$\frac{1}{\Delta t}(e^{v\Delta t} - \xi) = O(\Delta t + \Delta x).$$

and so the scheme is consistent and has "first-order accuracy." Hence the scheme is convergent if the von Neumann stability condition is satisfied. As we have seen in the previous section, this holds if

$$(2.29) \qquad\qquad r(2k + U\Delta x) \leq 1.$$

Note 1. In the usual terminology, the definition of "stability" is somewhat more complicated and less restrictive than the one introduced in §2.4. It has the advantage that with that definition a numerical scheme that is consistent is convergent *if and only if* it is stable. This is sometimes referred to as the *Lax equivalency theorem*.

Note 2. If we get stability *without* requiring any further relationship between Δx and Δt as they both approach zero, we say the finite difference scheme is *unconditionally* stable. If, on the other hand, we require that Δx and Δt approach zero in a specified manner (such as, for example, condition (2.29)) then we say the scheme is *conditionally stable*. In Chapter 4 we will encounter "implicit" schemes that although more complicated, have the advantage of being *unconditionally* stable.

PROBLEM 2.8.1. Show that the box scheme

$$\frac{1}{2k}[(u_m^{n+1} + u_{m+1}^{n+1}) - (u_m^n + u_{m+1}^n)] + \frac{a}{2h}[(u_{m+1}^{n+1} - u_m^{n+1}) + (u_{m+1}^n - u_m^n)] = 0$$

is consistent with the one-way wave equation $u_t + au_x = 0$ and find for which values of σ it is stable ($\sigma = ak/h$, $h = \Delta x$, $k = \Delta t$).

PROBLEM 2.8.2. Show that the box scheme

$$\frac{1}{2k}[(u_m^{n+1} + u_{m+1}^{n+1}) - (u_m^n + u_{m+1}^n)] + \frac{a}{2h}[(u_{m+1}^{n+1} - u_m^{n+1}) + (u_{m+1}^n - u_m^n)] = 0$$

is an approximation to the one-way wave equation $u_t + au_x = 0$ that is second-order accurate.

PROBLEM 2.8.3. Solve $u_t + u_x = 0$, $-1 \leq x \leq 1$, $0 \leq t \leq 1.2$ with $u(0, x) = \sin 2\pi x$ and periodic boundary conditions, i.e., $u(t, 1) = u(t, -1)$. In addition to finding the (simple) *exact* solution, use the following two finite difference schemes and compare each of the resulting solutions with the exact solutions.
(a) Forward in t, backward in x scheme with $\Delta t/\Delta x = 0.8$.
(b) Lax–Wendroff with $\Delta t/\Delta x = 0.8$.

Demonstrate numerically the first-order accuracy of the solution of (a) and the second-order accuracy of the solution of (b), using $h = 1/10, 1/20, 1/40$, and $1/80$. Do this by measuring the error e in the maximum norm and comparing $e/\Delta x$ and $e/(\Delta x^2)$ for succeeding small values of h. If these ratios eventually do not change appreciably, that indicates first- and second-order accuracy of the difference scheme.

2.9. Summary

We introduced two partial differential equations modeling the spread of impurities in the air: the advection equation and the advection diffusion equation. The method of characteristics can be used to solve the advection equation but is too cumbersome in real situations. It is often useful to use finite difference schemes provided one can assess the accuracy of the schemes.

One scheme (2.15) for the advection equation seems to work but has the disadvantage of introducing an artificial "dissipative" effect (i.e., the height of the "hump" seems to decrease with time), while it does seem to keep track of the location of the "hump."

The Lax–Wendroff scheme (Problem 2.7.6) does not seem to introduce any significant artificial dissipation but introduces some artificial oscillations instead, i.e., little "wiggles" seem to be superimposed on the graphs.

The von Neumann stability criterion was introduced and used to study a finite difference scheme for the advection diffusion equation. Numerical experiments were in excellent agreement with the von Neumann criterion. Two-and three-dimensional graphs for the solution of the advection diffusion equation as well as the pure advection equation have been obtained.

The relationship between stability, consistency, and convergence was investigated.

References

Air quality modeling is described in

[1] A. Friedman, *Mathematics in Industrial Problems, Part* 3, Chap. 11, IMA Vol. Math. Appl., Vol. 31, Springer-Verlag, New York, 1990, and the references therein.

Stability and convergence of difference schemes are discussed extensively in

[2] R. D. Richtmyer and K. W. Morton, *Difference Methods for Initial-Value Problems*, 2nd edition, Wiley–Interscience, New York, 1967.

[3] J. C. Strikwerda, *Finite Difference Schemes and Partial Differential Equations*, Wadsworth and Brooks/Cole Advanced Books and Software, Pacific Grove, CA, 1989.

Chapter 3

Electron Beam Lithography

3.1. Background

Semiconductor technology is playing an increasing role in instrumentation and sensor design. A semiconductor chip has typically a rectangular shape with edge size of 0.1 mm to 1 cm. It consists of a large number (on the order of one million) of tiny building blocks called devices. Each device is "doped" so that it functions as a (miniature) transistor. The devices have different shapes, and are interconnected by wires so as to form a Very Large Integrated Circuit (VLSI) that will perform certain tasks. The design of the chip, or of the VLSI, is done by electrical engineers. They use computer aided design to represent the geometric shapes of the pattern of devices to be created. The pattern needs to be imprinted on a silicon slab. The procedure is to first put resist coat on the silicon; this is some kind of a sensitive polymer. Then one uses the previously prepared computer programs for the design of the geometric shapes to guide an electron-beam (E-beam) so that it radiates photons only over certain areas (See Figs. 3.1 (a), (b)). After development of the exposed layer by means of a solvent sprayed over the resist coat (which "eats up" or "etches out" the areas exposed to the radiation by the E-beam), one obtains the desired pattern of windows and channels, as illustrated in Fig. 3.2. The miniature devices are built upon these curved out patterns.

What was described above is called *electron beam lithography*. There are other lithographic methods in use, such as those which employ optical beams (laser beams). The use of "masks" to guide the optical beam is quite common. The industry of solid state devices, and in particular of chips used in building computers, is a very fast growing industry, and new methods of production are continuously being developed.

In this chapter we consider the effectiveness of the E-beam method. Figure 3.3 shows the difference between the ideal situation (100% precision) and the imprecision under actual manufacturing conditions.

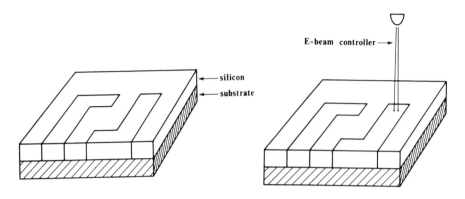

FIG. 3.1. (a) *The desired pattern.* (b) *Electron beam in action.*[1]

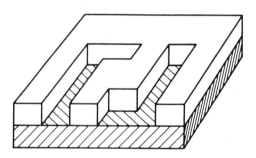

FIG. 3.2. *Imprinted pattern.*[2]

The causes for imprecision are the *forward scattering* of the electron beam hitting the surface and *backward scattering* of electrons as they bounce back from the substrate. Suppose the desired pattern is as in the grey shape in Fig. 3.4 (a). Because of the forward and backward scattering the actual shape may turn out to be as in Fig. 3.4 (b) and this may result in a faulty chip.

3.2. The mathematical model

Let us first assume only forward scattering. Suppose we apply dose $D(x)$ at each point $x = (x_1, x_2)$ in $\mathbb{R}^2$. Introduce the Gaussian kernel

$$(3.1) \qquad\qquad K_\alpha(x) = \frac{1}{\pi\alpha^2}\, e^{-|x|^2/\alpha^2}$$

[1]A. Friedman, *Mathematics in Industrial Problems*, IMA Vol. Math. Appl., Vol. 24, Springer-Verlag, New York, 1989, Figs. 9.1 and 9.3, pp. 80–81. ©1989, Springer-Verlag. Reprinted with permission.

[2]A. Friedman, *Mathematics in Industrial Problems*, IMA Vol. Math. Appl., Vol. 24, Springer-Verlag, New York, 1989, Fig. 9.2, p. 80. ©1989, Springer-Verlag. Reprinted with permission.

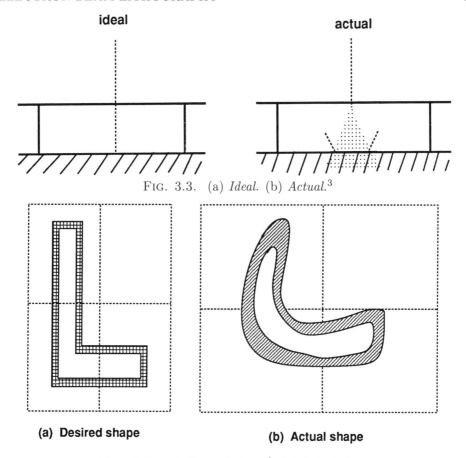

FIG. 3.3. (a) *Ideal.* (b) *Actual.*[3]

(a) Desired shape   **(b) Actual shape**

FIG. 3.4. (a) *Desired shape.*[4] (b) *Actual shape.*

with positive parameter α. Then as a result of the scattering, the actual exposure $E(x)$ at x is found experimentally to be a certain "average" of D, namely,

$$(3.2) \qquad\qquad E(x) = \int_{\mathbb{R}^2} K_\alpha(x - y) D(y) dy,$$

for some α which depends on the silicon material and on the E-beam device. Typically $\alpha \sim 0.1 - 0.2\mu m$.

If we denote by P the physical shape to be etched out (as in Fig. 3.2),

[3]A. Friedman, *Mathematics in Industrial Problems*, IMA Vol. Math. Appl., Vol. 24, Springer-Verlag, New York, 1989, Fig. 9.4, p. 82. ©1989, Springer-Verlag. Reprinted with permission.

[4]A. Friedman, *Mathematics in Industrial Problems*, IMA Vol. Math. Appl., Vol. 24, Springer-Verlag, New York, 1989, Fig. 9.5(c), p. 82. ©1989, Springer-Verlag. Reprinted with permission.

then the problem is

(3.3)   Find a function $D(x)$ such that $E(x) = \begin{cases} 1 & \text{if } x \in P \\ 0 & \text{if } x \notin P. \end{cases}$

If we also assume backward scattering with parameter β then the experimental formula for E is revised to

(3.4) $E(x) = \dfrac{1}{1+\eta} \displaystyle\int_{\mathbb{R}^2} [K_\alpha(x-y) + \eta K_\beta(x-y)]D(y)dy,$

where η is the ratio of the intensity of backward to forward scattering, and $\beta \sim 2 - 3\mu m, \ 0 < \eta < 1$.

Here again the basic problem is (3.3).

Equations (3.2), (3.4) are called the *proximity effect equations*.

3.3. The heat equation

Let us concentrate first on (3.2).

For any bounded function $f(x)$, define

(3.5) $u(x,t) = \displaystyle\int_{\mathbb{R}^2} \dfrac{1}{4\pi t} \exp\left(-\dfrac{|x-y|^2}{4t}\right) f(y)dy.$

Then

(3.6) $E(x) = u(x,t) \quad \text{if } t = \dfrac{\alpha^2}{4}, \ f(y) \equiv D(y),$

and problem (3.3) becomes

(3.7) $u(x,t) = \chi_P(x) \quad \text{at } t = \dfrac{\alpha^2}{4},$

where χ_P is the characteristic function of P. $(\chi_P(x) = 1$ if $x \in P, \ \chi_P(x) = 0$ if $x \notin P.)$

The integral in (3.5) is convergent if $f(y)$ is bounded function; this will always be assumed. We will also assume that $f(y)$ is continuous.

PROBLEM 3.3.1. Prove that the function u given by (3.5) satisfies

(3.8) $\dfrac{\partial u}{\partial t} = \dfrac{\partial^2 u}{\partial x_1^2} + \dfrac{\partial^2 u}{\partial x_2^2}.$

Equation (3.8) is called the *heat equation*; the temperature $u(x,t)$ within a body D in $\mathbb{R}^n$ is known to satisfy the heat equation

(3.9) $u_t = \Delta u \quad \text{where } \Delta u = \displaystyle\sum_{i=1}^{n} \dfrac{\partial^2 u}{\partial x_i^2}$

(here $x = (x_1, \ldots, x_n)$). For any $n \geq 1$, the function

$$(3.10) \qquad u(x,t) = \int_{\mathbb{R}^n} \frac{1}{(4\pi t)^{n/2}} \exp\left(-\frac{|x-y|^2}{4t}\right) f(y) dy$$

satisfies the heat equation (3.9).

THEOREM 3.3.1. *The function u given in (3.10) also satisfies*

$$\lim_{t \to 0} u(x,t) = f(x).$$

Proof. We give the proof for $n = 1$ and leave the generalization to any n to the reader. We set

$$K(x,y,t) = (4\pi t)^{-1/2} e^{-|x-y|^2/4t}.$$

Letting $y = x + (4t)^{1/2}\eta$, we obtain

$$(*) \qquad \int_{|y-x|>\delta} K(x,y,t) dy = \frac{1}{\sqrt{\pi}} \int_{|\eta|>\delta/\sqrt{4t}} e^{-|\eta|^2} d\eta.$$

Setting $\delta = 0$ we get

$$\int_{-\infty}^{\infty} K(x,y,t) dy = 1 \quad \text{for } t > 0,$$

where we have used the well-known identity

$$\int_{-\infty}^{\infty} e^{-\eta^2} d\eta = \sqrt{\pi}.$$

By the boundedness of f we may write $|f(x)| \leq M$. By the continuity of f, given any $\varepsilon > 0$, we can find a positive δ such that

$$|f(y) - f(\xi)| < \varepsilon \quad \text{for } |y - \xi| < 2\delta.$$

Then, for $|x - \xi| < \delta$,

$$|u(x,t) - f(\xi)| = |\int K(x,y,t)(f(y) - f(\xi)) dy|$$

$$\leq \int_{|y-x|<\delta} K(x,y,t)|f(y) - f(\xi)| dy$$

$$+ \int\limits_{|y-x|>\delta} K(x,y,t)|f(y) - f(\xi)|dy$$

$$\leq \int\limits_{|y-\xi|<2\delta} K(x,y,t)|f(y) - f(\xi)|dy + 2M \int\limits_{|y-x|>\delta} K(x,y,t)dy$$

$$\leq \varepsilon \int\limits_{-\infty}^{\infty} K(x,y,t)dy + 2M \int\limits_{|y-x|>\delta} K(x,y,t)dy.$$

The first term equals ε, while the second term can be made less than ε for t sufficiently small, using $(*)$, and the fact that

$$\int\limits_{-\infty}^{\infty} e^{-\eta^2} d\eta$$

is a convergent improper integral.

Consider the problem of solving

$$(3.11) \qquad \frac{\partial u}{\partial t} = \Delta u \quad \text{for } x \in \mathbb{R}^n, \quad t > 0,$$

subject to the initial condition

$$(3.12) \qquad u(x,0) = f(x) \quad \text{for } x \in \mathbb{R}^n.$$

This is called a *Cauchy problem*.

We have shown that the expression in (3.10) is a solution of this problem.

THEOREM 3.3.2. *There is only one bounded solution to the Cauchy problem* (3.11), (3.12).

Proof. Suppose u_1, u_2 are two bounded solutions and set $u = u_1 - u_2$. Then u solves the heat equation and $u(x,0) = 0$. Consider the function

$$(3.13) \qquad w = u + \varepsilon \left(\frac{|x|^2}{2n} + 2t \right), \qquad \varepsilon > 0.$$

We claim that, for any $T > 0$, $w(x,t) \geq 0$ if $x \in \mathbb{R}^n$, $0 \leq t \leq T$. Indeed, since $w(x,0) \geq 0$ and $w(x,t) \to \infty$ if $|x| \to \infty$, if the assertion is not true then w takes negative values and, in fact, has a negative minimum at some point (x_0, t_0) with $0 < t_0 \leq T$. At that point

$$\frac{\partial w}{\partial t} \leq 0 \quad \text{and} \quad -\Delta w \leq 0$$

so that $w_t - \Delta w \leq 0$. Since, however, by (3.13),

$$w_t - \Delta w = \varepsilon > 0,$$

this is a contradiction. Having proved that $w \geq 0$ we now let $\varepsilon \to 0$ and conclude that $u \geq 0$. Similarly we prove that $-u \geq 0$, so that $u_1 - u_2 \equiv u \equiv 0$.

PROBLEM 3.3.2. Let u satisfy $u_t - \Delta u \geq 0$ in a cylinder $D \times (0, T)$, D a bounded domain in $\mathbb{R}^n$. Assume that u is continuous in the closure $\overline{D} \times [0, T]$ and $u(x, 0) \geq 0$ $(x \in \overline{D})$, $u(x, t) \geq 0$ if $x \in \partial D$, $0 \leq t \leq T$. Prove that $u \geq 0$ in $D \times (0, T)$.

This result is often called the *weak maximum principle*. It actually states that u cannot take on inside the domain a minimum value which is strictly smaller than all its boundary values on $\overline{D} \times \{0\}$ and on $\partial D \times \{0 \leq t \leq T\}$.

PROBLEM 3.3.3. Solve the Cauchy problem $(0 < t < 10)$

$$u_t = u_{xx},$$
$$u(x, 0) = \chi_{(-1,1)}$$

by finite differences using the "Euler forward scheme,"

$$\frac{u_j^{n+1} - u_j^n}{\Delta t} = \frac{u_{j+1}^n - 2u_j^n + u_{j-1}^n}{(\Delta x)^2} .$$

Use the von Neumann criterion to discuss the stability of this scheme (see Chapter 2).

3.4. The proximity effect

We can now put problem (3.3) in appropriate context. Indeed that problem can be formulated as follows.

Find initial values f such that if we solve (3.11), (3.12) for time $0 < t \leq T$ where $T = \alpha^2/4$, then

(3.14) $$u(x, T) = \chi_P(x).$$

In other words, given *terminal* values $u(x, T)$, as in (3.14), solve the heat equation *backward* in time for $t \leq T$ and find the values $u(x, 0)$. These values will give us the dose needed in (3.3).

This problem is often called the *backward heat problem*. It is "ill-posed" because, as we shall see in the next section, the solution does not always exist.

The maximum principle (discussed earlier) provides some useful insight. If the dose is increased from $u(x, 0)$ to $v(x, 0)$ (i.e., $v(x, 0) \geq u(x, 0)$) then, by the maximum principle, the corresponding solutions $v(x, t)$ and $u(x, t)$ of the heat equation satisfy $v(x, T) \geq u(x, T)$. Thus, if the dose is increased, so is the exposure. Since zero dose gives zero exposure and since we cannot have a negative dose, this also shows that the exposure cannot be negative.

3.5. The use of Fourier series

If we apply a dose evenly distributed over a pattern in the form of the letter
"L," for example, with or without back scattering, we actually obtain a
somewhat "rounded out" version of that pattern (see the computer generated
Fig. 3.5).

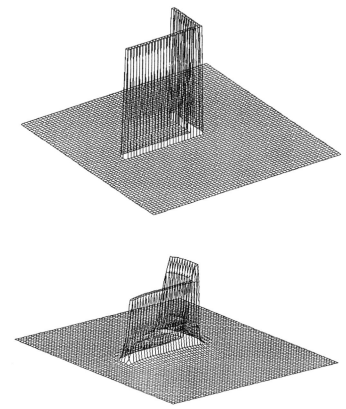

FIG. 3.5. (a) *Dose $D(x)$ evenly applied to figure "L."* (b) *Resulting $E(x)$ with
back-scattering $\alpha = 1$, $\beta = 10$, $\eta = \frac{1}{2}$. Gaussian kernel truncated: $|x| \leq 200$.*

PROBLEM 3.5.1. Try this with other letters.

The above are computations of the direct problem. Now we want to study
the real-life problem that is the inverse problem. For simplicity, we concentrate
on the one-dimensional case, with the dose being given by the characteristic
function of the interval $|x| \leq \frac{1}{2}$, say. The resulting exposure is described in
Fig. 3.6. One approach to improve the exposure is by guesswork and trial and
error. For example, we might be tempted to try a dose of the form shown in
Fig. 3.7, and see what we obtain for the resulting exposure.

To be able to make an educated guess, we are going to use the method

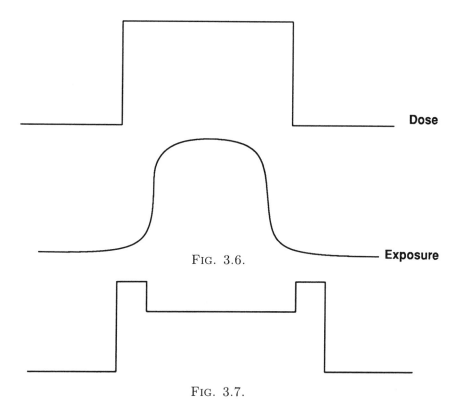

FIG. 3.6.

Dose

Exposure

FIG. 3.7.

of Fourier series and separation of variables to try to solve the forward and backward heat equation (approximately) even with back-scattering.

Some introductory words on the method of separation of variables to solve partial differential equations are in order.

Let us begin by recalling that in solving a homogeneous *ordinary* differential equation with constant coefficient, say the equation

$$y'' - y = 0 \quad \text{with } y(0) = 1 \quad \text{and} \quad y'(0) = 0,$$

we look for solutions of a certain form: $y = e^{mx}$. Substituting, we see that possible solutions are e^x and e^{-x}. We then try to use these special solutions as *building blocks* to construct a solution satisfying the initial conditions. Thus substituting the trial solution $y = Ae^x + Be^{-x}$ into the homogeneous initial condition $y'(0) = 0$ we obtain $A - B = 0$. Substituting $y = A(e^x + e^{-x})$ into the remaining nonhomogeneous condition $y(0) = 1$, we get $A = \frac{1}{2}$, thus giving us the unique solution $y(x) = \frac{1}{2}(e^x + e^{-x}) = \cosh x$.

To be able to use Fourier series we must replace the "initial" function

$$f(x) = \begin{cases} 1 & \text{for } |x| < \frac{1}{2} \\ 0 & \text{otherwise} \end{cases}$$

by its periodic counterpart. For example, let $f(x)$ be defined as above in the interval $(-\pi, \pi)$ and be repeated periodically, to obtain the "square wave" function pictured in Fig. 3.8.

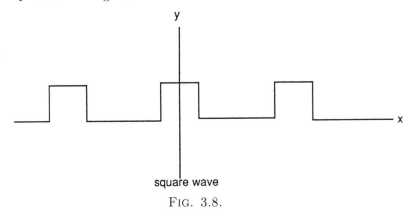

square wave

FIG. 3.8.

We then look for a solution to the heat equation

$$u_t = u_{xx}$$

of the form $X(x)T(t)$, where $X(x)$ is an even periodic function of period 2π. Substituting in the heat equation, we obtain $X(x)T'(t) = X''(x)T(t)$, or

$$\frac{T'(t)}{T(t)} = \frac{X''(x)}{X(x)} \quad \text{for all } x \text{ and positive } t.$$

Both sides must be equal to the same constant λ, hence $X''(x) = \lambda X(x)$ for all x. The only even periodic solutions (of period 2π) are obtained when $\lambda = -n^2$, $n = 0, 1, 2, 3 \ldots$ in which case we get the family of solutions $\cos nx$. The corresponding $T(t)$ must be a multiple of $e^{-n^2 t}$, thus giving us a family of "separated" solutions

$$e^{-n^2 t} \cos nx.$$

We now try to use these as building blocks to satisfy the remaining (nonhomogeneous) initial condition, just as in the case of the ordinary differential equation. Thus we try (in the traditional notation)

$$u(x, t) = \frac{a_0}{2} + \sum_{1}^{\infty} a_n e^{-n^2 t} \cos nx,$$

and impose the initial condition

$$u(x, 0) = \frac{a_0}{2} + \sum_{1}^{\infty} a_n \cos nx = f(x).$$

By Fourier's theorem every even periodic function satisfying some mild restrictions can be expanded as such a cosine series, with coefficients a_n given by

$$a_n = \frac{2}{\pi} \int_0^\pi f(x) \cos nx \, dx.$$

Under these "mild" conditions on $f(x)$ the series for $u(x,t)$ actually converges and solves the initial value problem for the heat equation, provided f is periodic and even.

Note that the Fourier (cosine) coefficients of the function $u(x,0)$ are the numbers a_n, while the Fourier coefficients of $u(x,t)$ are the numbers

$$a_n e^{-n^2 t}.$$

Thus to get from the coefficients at time t_1 to the coefficients at time t_2 multiply the nth coefficient by the number $e^{-n^2(t_2-t_1)}$.

At first glance this argument would seem to give us a method for solving the backward heat equation, i.e., prescribe the square wave function $f(x)$ at a time $t = T > 0$ and then try to evaluate

$$u(x,0) = \frac{1}{2} a_0 + \sum_1^\infty a_n e^{n^2 T} \cos nx.$$

The trouble, of course, is that this series will in general not converge, and hence that a solution to the "backward heat problem" does not exist in general. Thus the backward heat problem is not *well posed* (i.e., it is *ill posed*). In general terms, we say that a problem is *well posed* if it always has exactly *one* solution. (Sometimes additional conditions are imposed, but we need not worry about these now.)

To give us some idea of the dose $D(x)$ required to achieve a desired exposure $E(x)$ (say the square wave function considered above), we truncate $E(x)$ by considering only the first N terms (say 10) in its cosine series expansion

$$E(x) = \frac{1}{2} e_0 + \sum_1^\infty e_n \cos nx,$$

i.e.,

$$E_N(x) = \frac{1}{2} e_0 + \sum_1^N e_n \cos nx$$

and consider the "approximate dose"

$$D_N(x) = \frac{1}{2} d_0 + \sum_1^N d_n \cos nx,$$

where $d_0 = e_0$ and $d_n = e_n e^{n^2 t}$,

$$D_N(x) = \frac{1}{2} e_0 + \sum_1^N e_n e^{n^2 t} \cos nx$$

(t is considered fixed here).

Several comments are in order: The series for $D_N(x)$ converges, since there are only a finite number of terms. Second, as was pointed out in previous sections, the "heat equation" that is being studied here has nothing to do with the flow of heat! "t" is only a convenient parameter related to scattering (and later backward scattering). Finally, there is a chance that by the right choice of N, and t, the graph of D_N will give us a clue as to how to modify the dose to get an exposure close to the desired one.

3.6. The inclusion of backward scattering

We noted that formulas (3.2) (and (3.4)) could be related to the corresponding formulas of type (3.5) provided we use the relation $t = \alpha^2/4$. Suppose we agree to the notation $t_1 = \alpha^2/4$, $t_2 = \beta^2/4$. Then (3.4) essentially says the following: Start with $D(x)$ as initial data at $t = 0$; solve the (forward) heat equation and evaluate at time t_1, and multiply the result by $1/(1 + \eta)$. Start afresh with $D(x)$ again, but this time evaluate the solution at $t = t_2$, and multiply the result by $\eta/(1 + \eta)$. Add the two functions of x. The result will be $E(x)$.

Now the whole procedure can be carried out in terms of Fourier series. However, to make the process converge, we work with $D_N(x)$ and $E_N(x)$ instead.

The result is

$$e_n = \frac{1}{1 + \eta} d_n e^{-n^2 t_1} + \frac{\eta}{1 + \eta} d_n e^{-n^2 t_2}$$

$$= \left(\frac{e^{-n^2 t_1}}{1 + \eta} + \frac{\eta}{1 + \eta} e^{-n^2 t_2} \right) d_n$$

or

$$d_n = \frac{1 + \eta}{e^{-n^2 t_1} + \eta e^{-n^2 t_2}} e_n.$$

Thus the procedure is

For $0 \leq n \leq N$ calculate the e_n by

$$e_n = \frac{2}{\pi} \int_0^\pi E(x) \cos nx \, dx.$$

Calculate the d_n by the above-mentioned formula. Then calculate

$$D_N(x) = \frac{1}{2} d_0 + \sum_1^N d_n \cos nx .$$

3.7. Computational experiments

Figure 3.9 shows the results of this procedure graphically for the square wave function that equals one in $|x| \leq \frac{1}{2}$. A constant term of 0.5 has been added to avoid negative values. However this does not satisfactorily solve the problem of negative values arising presumably because of truncation of the Fourier series.

In Fig. 3.10 a similar procedure is exhibited, this time for the square wave of the function that equals one in $|x| \leq 1$. Here the graph of E_{10} is also exhibited. Figure 3.11 is similar.

PROBLEM 3.7.1. Try varying the values of n, η, t_1, and t_2.

PROBLEM 3.7.2. Generalize the method of Fourier series described above to the case where $E(x)$ is periodic *but not even*.
Hint. Use Fourier's theorem for general periodic functions.

PROBLEM 3.7.3. Try the method of "experimentation and guesswork" described by Fig. 3.7.

3.8. Summability of Fourier series

In the last section Fourier series methods were introduced to solve the inverse problem approximately. The results were graphed and seem to give some information for small values of α, β, η but not for others. Oscillations, to the extent that they exist, indicate a weakness of the method. Another weakness is the appearance of negative values where they do not make sense physically. This is no doubt due to the fact that when we approximate a nonnegative function by a partial sum of its Fourier series, that partial sum may be negative in places.

The above-mentioned difficulty can be met by replacing the ordinary notion of *convergence* of a Fourier series by that of *summability* (or "Cesaro summability") of a Fourier series (or series in general). Thus if (in the usual notation)

$$s_n(x) = \frac{a_0}{2} + \sum_{\nu=1}^n (a_\nu \cos \nu x + b_\nu \sin \nu x) \equiv \sum_{\nu=0}^n u_\nu$$

represents the "nth" partial sum of the Fourier series of a function $f(x)$, then the *average* of the first n partial sums

$$S_n = \frac{1}{n}(s_0 + s_1 + s_2 + \cdots s_{n-1})$$

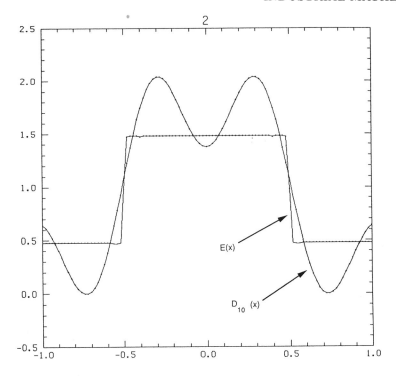

FIG. 3.9. $\eta = \frac{1}{4}, t_1 = 1/100, t_2 = 1$.

$$= \frac{1}{n} \sum_{\nu=0}^{n} (n - \nu) u_\nu = \sum_{\nu=0}^{n} \left(1 - \frac{\nu}{n} \right) u_\nu$$

represents *another* approximation to the function $f(x)$. Here

$$u_\nu = a_\nu \cos \nu x + b_\nu \sin \nu x \quad \text{if} \quad \nu > 0 \quad \text{and} \quad u_0 = \frac{a_0}{2}.$$

This will have the desirable property of *preserving* positivity: if $f(x) \geq 0$ then $S_n(x) \geq 0$.

That $S_n(x)$ is an approximation to $F(x)$ is expressed by the following theorem.

FÉJER'S THEOREM. *If $S_n(x)$ represents the average of the first n partial sums of the Fourier series of a continuous function of period 2π, then $S_n(x)$ converges uniformly for all x to $f(x)$.*

This means that *given any positive number ε, there exists an $N = N(\varepsilon)$ such that for all $n > N$*

$$|f(x) - S_n(x)| < \varepsilon \quad \text{for all} \quad x.$$

Féjer's theorem and the positivity preserving property of the partial sums are established in the Appendix to this chapter.

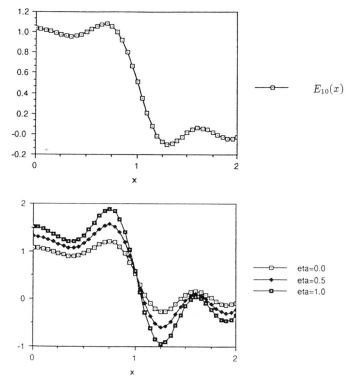

FIG. 3.10.   (a) $E_{10}(x)$. (b) $D_{10}(x)$ *for* $t_1 = 0.1, t_2 = 1.0$ *for various eta's.*

In the case where $f(x)$ has an isolated jump discontinuity at a point x_0, $S_n(x)$ will converge to the average of the right- and left-hand limits of $f(x)$ at x_0, but the convergence will not be uniform.

Féjer's theorem is often expressed by saying that the Fourier series of a continuous function is *summable* to the function.

Let us call $S_n(x)$ the nth *Féjer sum* of the function $f(x)$.

EXERCISE 3.8.1. Repeat the computations of §3.7, but instead of the partial sums E_N and D_N use the corresponding Féjer sums $\widetilde{E}_N$ and $\widetilde{D}_N$. Compare the results; see also Figs. 3.12 and 3.13.

EXERCISE 3.8.2. Try to improve the results of the previous exercise by "numerical experimentation."

3.9. Summary

We have introduced a mathematical model for electron beam lithography using an integral transform with a Gaussian kernel, including both forward and backward scattering. The relationship to the "heat equation" was discussed. Fourier series methods were introduced to solve the inverse problem

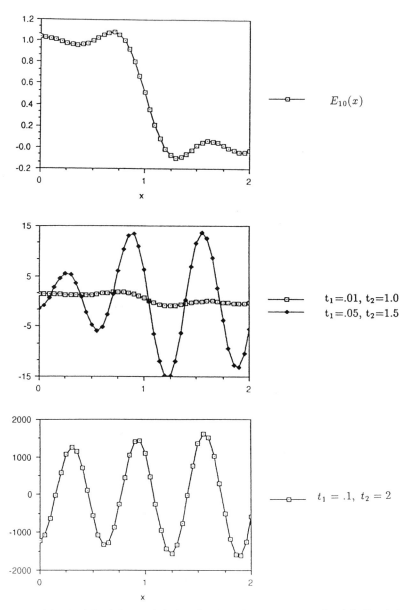

FIG. 3.11. (a) $E_{10}(x)$. (b) $D_{10}(x)$ for fixed eta's at various t's. (c) $D_{10}(x)$ for $eta = 1.0$ at $t_1 = 0.1, t_2 = 2.0$.

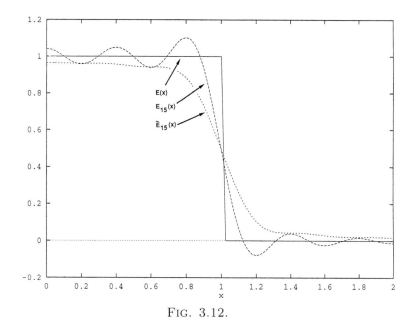

FIG. 3.12.

approximately. The results are graphed and seem to give some information for some values of α, β, η, but not for others. Oscillations, to the extent that they exist, are taken to indicate a weakness of the method.

A second method based on Féjer's theorem was introduced. Partial sums were replaced by Féjer sums. Oscillations were damped and negative values were eliminated in some cases, diminished in others. It might be possible to further improve on these results by "numerical experimentation."

3.10. Appendix: Proof of Féjer's theorem

Recall the definition of (Cesaro) summability.[5] Let $\sum_0^\infty u_n$ be a given infinite series and let $s_n = \sum_{\nu=0}^n u_\nu$ be the nth partial sum of the series. If the average of the first n terms is formed

$$S_n(x) = \frac{1}{n}(s_0 + s_1 + \cdots + s_{n-1}) = \frac{1}{n}\sum_{\nu=0}^n (n-\nu)u_\nu = \sum_{\nu=0}^n \left(1 - \frac{\nu}{n}\right) u_\nu$$

then the given sum is said to be *summable with sum U*, if $S_n \to U$ for $n \to \infty$.

Note. If the series $\sum_0^\infty u_n$ is convergent with sum U, then it is also summable with the same sum U. On the other hand, there exist series that are summable but not convergent.

The Féjer theorem states the following.

<hr />

[5]We follow the treatment in H. Bohr, *Almost Periodic Functions*, Chelsea, 1947. ©1947, Chelsea Publishing Company.

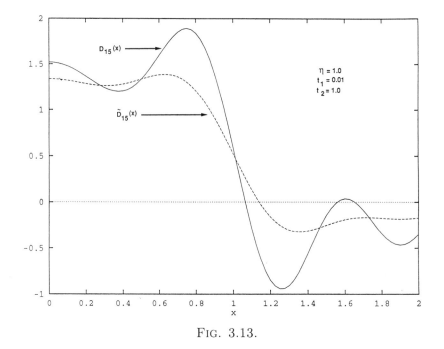

FIG. 3.13.

The Fourier series

$$\sum_0^\infty u_n = a_0 + (a_1 e^{ix} + a_{-1} e^{-ix}) + \cdots + (a_n e^{inx} + a_{-n} e^{-inx}) + \cdots$$

of a continuous function $f(x)$ of period 2π is always summable, and, moreover, uniformly summable, with sum $f(x)$, i.e.,

$$S_n(x) = \frac{1}{n}(s_0(x) + s_1(x) + \cdots + s_{n-1}(x)) = \frac{1}{n} \sum_{\nu=-n}^n (n - |\nu|) a_\nu e^{i\nu x}$$

$$= \sum_{\nu=-n}^n \left(1 - \frac{|\nu|}{n}\right) a_\nu e^{i\nu x}$$

approaches $f(x)$ uniformly for all x as $n \to \infty$. In other words, we have that $max|S_n(x) - f(x)| \to 0$ as $n \to \infty$.

The proof rests on the transformation of the expression for $S_n(x)$. Since for a fixed value of x

$$f(x + t) \quad \text{has the Fourier series} \quad \Sigma a_\nu e^{i\nu x} \cdot e^{i\nu t},$$

we have first for the nth partial sum

$$s_n(x) = \sum_{\nu=-n}^n a_\nu e^{i\nu x} = \sum_{\nu=-n}^n \int \{f(x + t)e^{-i\nu t}\} \frac{dt}{2\pi}$$

$$= \int \left\{ f(x+t) \sum_{\nu=-n}^{n} e^{-i\nu t} \right\} \frac{dt}{2\pi} = \int \{ f(x+t) D_n(t) \} \frac{dt}{2\pi},$$

where the Dirichlet kernel $D_n(t)$ is given by

$$D_n(t) = \sum_{\nu=-n}^{n} e^{-i\nu t} = \frac{e^{i(n+1)t} - e^{-i(n+1)t}}{e^{it/2} - e^{-it/2}} = \frac{\sin\left(n + \frac{1}{2}\right)t}{\sin\frac{t}{2}}$$

$$= \frac{\cos nt - \cos(n+1)t}{2\sin^2\frac{t}{2}}$$

and integrals without limits mean integrals from 0 to 2π. For the average value $S_n(x)$, we find a corresponding expression

$$S_n(x) = \sum_{\nu=-n}^{n} \left(1 - \frac{|\nu|}{n}\right) a_\nu e^{i\nu x} = \int \{f(x+t) K_n(t)\} \frac{dt}{2\pi},$$

where $K_n(t)$ is the Féjer kernel given by

$$K_n(t) = \sum_{\nu=-n}^{n} \left(1 - \frac{|\nu|}{n}\right) e^{-i\nu t} = \frac{1}{n} \sum_{\nu=-n}^{n} (n - |\nu|) e^{-i\nu t}$$

$$= \frac{1}{n} (D_0(t) + D_1(t) + \cdots + D_{n-1}(t))$$

$$= \frac{1}{n} \frac{1 - \cos nt}{2\sin^2\frac{t}{2}} = \frac{1}{n} \left(\frac{\sin\frac{nt}{2}}{\sin\frac{t}{2}}\right)^2.$$

Of these expressions for $K_n(t)$ the first and (particularly) the last are important. From the first it follows that

$$\int \{K_n(t)\} \frac{dt}{2\pi} = 1$$

(since the constant term is 1) and from the last that (unlike $D_n(t)$)

$$K_n(t) \geq 0$$

for all t. Figure 3.14 shows the graph of $K_n(t)$. (If t is a multiple of 2π, then $K_n(t) = n$; if however, nt, but not t, is a multiple of 2π, then $K_n(t) = 0$). Now let $\varepsilon > 0$ be given an arbitrary value. We have to show that

$$|S_n(x) - f(x)| < \varepsilon$$

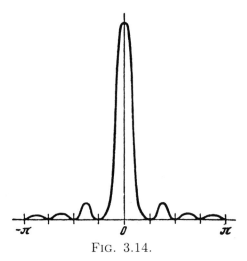

$$\text{Fig. 3.14.}$$

for all x, when $n > N = N(\varepsilon)$. For an arbitrary fixed x

$$S_n(x) - f(x) = \int \{f(x+t)K_n(t)\}\frac{dt}{2\pi} - f(x)\int \{K_n(t)\}\frac{dt}{2\pi}$$

$$= \int \{(f(x+t) - f(x))K_n(t)\}\frac{dt}{2\pi}.$$

Now $f(x)$ is continuous and periodic, hence $f(x)$ is bounded and uniformly continuous, i.e.,

$$|f(x)| \leq C$$

for all x, and, for any $\eta > 0$, and $|f(x_1) - f(x_2)| \leq \eta$ for $|x_1 - x_2| \leq \delta = \delta(\eta)$. We denote by $\omega(\delta)$ the maximum of $|f(x_1) - f(x_2)|$ for $|x_1 - x_2| \leq \delta$, and then $\omega(\delta) \to 0$ for $\delta \to 0$. We make use of the specified properties of $K_n(t)$ and obtain for an arbitrary δ $(< \pi)$

$$|S_n(x) - f(x)| \leq \frac{1}{2\pi} \int\limits_{-\pi}^{\pi} |f(x+t) - f(x)|K_n(t)dt$$

$$= \frac{1}{2\pi} \left[\int\limits_{-\delta}^{\delta} + \int\limits_{-\pi}^{-\delta} + \int\limits_{\delta}^{\pi} \right] |f(x+t) - f(x)|K_n(t)dt$$

$$\leq \frac{1}{2\pi} \left[\omega(\delta) \int\limits_{-\delta}^{\delta} K_n(t)dt + 2(\pi - \delta)2C \cdot \frac{1}{n}\frac{1}{\sin^2 \frac{\delta}{2}} \right]$$

$$< \frac{1}{2\pi} \left[\omega(\delta) \int_{-\pi}^{\pi} K_n(t)dt + 2\pi \frac{1}{n} \frac{2C}{\sin^2 \frac{\delta}{2}} \right]$$

$$= \omega(\delta) + \frac{1}{n} \frac{2C}{\sin^2 \frac{\delta}{2}}.$$

We first choose a *fixed* δ so small that $\omega(\delta) < \varepsilon/2$, and then an N so large that

$$\frac{1}{N} \frac{2C}{\sin^2 \frac{\delta}{2}} < \frac{\varepsilon}{2}.$$

Then for $n > N$ (and all x)

$$|S_n(x) - f(x)| < \frac{\varepsilon}{2} + \frac{\varepsilon}{2} = \varepsilon,$$

which proves the theorem.

Note. A change of variable $2\pi/\alpha \, x = x'$ extends the theorem to functions periodic of period α.

References

The forward and backward scattering was modeled and studied in

[1] J. M. Pavkovich, *Proximity effect correction calculation*, J. Vacuum Sci. Technol., B 4 (1986), pp. 159–163.

[2] P. D. Gerber, *Exact solution of the proximity effect equation by splitting method*, J. Vacuum Sci. Technol., B 6 (1988), pp. 432–435.

See also

[3] A. Friedman, *Mathematics in Industrial Problems, Part* 2, Chap. 9, IMA Vol. Math. Appl., Vol. 24, Springer-Verlag, New York, 1989.

Chapter 4

Development of Color Film Negative

4.1. The process

Film development is of major interest to many industries, yet the modeling of the process is not fully understood. Here we consider the development of a color film negative. Such a film consists of several photographic emulsion layers sandwiched between layers of gelatin. The thickness of each layer is approximately $15\mu m - 20\mu m$. The photographic emulsion consists of silver halide grains and oil droplets suspended in gelatin (the silver halide is a compound of silver with chlorine, bromine, or iodine). The size of an oil droplet is about $0.05\mu m$ across and of the silver grain is about $0.5\mu m$ across. The relative sizes are depicted in Fig. 4.1.

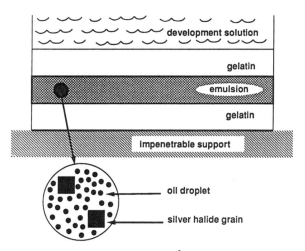

FIG. 4.1. [1]

[1] A. Friedman, *Mathematics in Industrial Problems*, IMA Vol. Math. Appl., Vol. 24, Springer-Verlag, New York, 1989, Fig. 10.1, p. 89. ©1989 Springer-Verlag. Reprinted with permission.

Oil droplets contain dye-forming coupler and inhibitor-forming coupler. (A *coupler* is a substance enhancing the formation of certain compounds.) Figure 4.1 shows the film (consisting, for simplicity, of only three layers) in contact with a well-stirred tank that contains the development solution. When the picture was taken, the film had been exposed to light and, as a result, silver grains that were exposed to light had acquired *latent image sites* ("latent" because they were so far still invisible). In the development process chemicals from the solution in the tank diffuse into the film and react with the exposed silver halide grains to start a sequence of chemical reactions that will form an image. A *reduced developer* R from the development solution begins the process by depositing electrons e^- onto the latent image sites.

There are always some loose positively charged silver ions, known as *interstitial silver ions*, wandering around inside the silver halide grain. They are attracted to the electrons e^- at the latent image site and combine with them to form pure silver. The pure silver grows into a silver filament and acts as a wire. The energetic circumstances created by the flow of interstitial ions dislodge additional silver ions at *kink sites*; a kink site is a location on the silver halide crystal lattice at which silver ions can easily be dislodged. As the process continues the silver halide grain is shrinking (new kink sites are formed while the grain is shrinking), and the pure silver filament is growing. A developing silver halide grain may give rise simultaneously to several filaments. Figure 4.2 summarizes the reaction of R with the exposed silver halide.

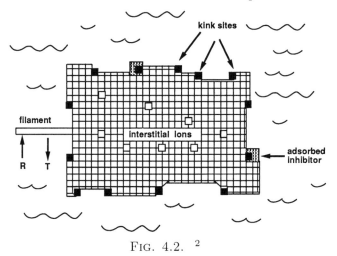

FIG. 4.2. [2]

[2]A. Friedman, *Mathematics in Industrial Problems*, IMA Vol. Math. Appl., Vol. 24, Springer-Verlag, New York, 1989, Fig. 10.2, p. 90. ©1989 Springer-Verlag. Reprinted with permission.

The reduced developer R, after reacting with the exposed silver halide grains (i.e., after giving up its electrons), becomes *oxidized developer* T. It then diffuses and reacts with a coupler C in the oil droplet to form a *dye* (which is immobile) and an *inhibitor* P. The inhibitor P diffuses and some of it adsorbs to the surface of the silver grain blocking halides from dissociating (see Fig. 4.2).

Let us denote by P^* the concentration of the adsorbed inhibitor on the surface of the silver halide grain. Figure 4.3 gives schematically the chemical processes described above.

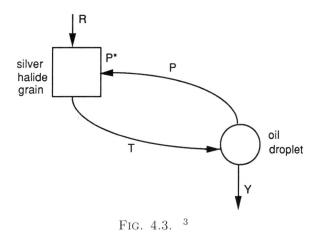

FIG. 4.3. [3]

4.2. The mathematical model

In the model to be described below we do not deal with the microscopic picture. In fact, we "smear out" the effect of individual grains and thus deal with a continuous model whereby $R, T, C, D, P,$ and P^* are density functions:

$$R = \text{reduced developer,}$$

$$T = \text{oxidized developer,}$$

$$C = \text{coupler (in the oil),}$$

$$Y = \text{dye,}$$

$$P = \text{inhibitor,}$$

$$P^* = \text{adsorbed inhibitor.}$$

[3]A. Friedman, *Mathematics in Industrial Problems*, IMA Vol. Math. Appl., Vol. 24, Springer-Verlag, New York, 1989, Fig. 10.3, p. 90. ©1989 Springer-Verlag. Reprinted with permission.

We also need to introduce a macroscopic version of

$$S = \quad \text{surface area of the silver halide grains per unit volume.}$$

Before writing the differential equations for these quantities, we explain the general modeling of a chemical concentration W in a solution undergoing both dynamical changes and chemical changes.

The dynamics usually consists of *diffusion* and *drift* terms. (Recall Chapter 2.) The diffusion is a local process whereby the concentration at each point tends to go up or down, to the level of the nearby concentrations. This causes a change in W quantified by

$$\nabla(D\nabla W),$$

where D is the *diffusion matrix*; D is positive definite symmetric matrix. In many cases D is actually a scalar. If D is a constant scalar then write the diffusion term as $D\nabla^2 W$ or $D\Delta W$.

The drift is a result of the fluid motion of the solution. In simple cases the drift is linear, and it is quantified by a term

$$B \cdot \nabla W \quad \text{or} \quad \Sigma b_i \frac{\partial W}{\partial x_i} \qquad (B = (b_1, b_2, b_3)),$$

where B is a function of space and time. The conservation of mass (or the equation of continuity) dictates that

$$(4.1) \qquad \frac{\partial W}{\partial t} = \nabla \cdot (D\nabla W) + B \cdot \nabla W.$$

Suppose now that another chemical V is present in the solution and that it interacts with W to produce a chemical U at a reaction rate k. Thus W decreases at a rate kVW so that (4.1) must be replaced by

$$(4.2) \qquad \frac{\partial W}{\partial t} = \nabla \cdot (D\nabla W) + B \cdot \nabla W - kVW.$$

If W is immobile, that is, no diffusion or drift take place, then (4.2) is replaced by

$$(4.3) \qquad \frac{\partial W}{\partial t} = -kVW.$$

The discussion above explains how the previous description of the development process leads to the following macroscopic model:

$$(4.4) \qquad \frac{\partial R}{\partial t} = D_R \Delta R - f_{\text{dev}}(R, P, S, P^*)E(x) \qquad (x = (x_1, x_2, x_3)),$$

$$(4.5) \qquad \frac{\partial T}{\partial t} = D_T \Delta T + f_{\text{dev}}(R, P, S, P^*)E(x) - k_1 TC,$$

(4.6) $\dfrac{\partial C}{\partial t} = -k_1 TC,$

(4.7) $\dfrac{\partial Y}{\partial t} = k_1 TC,$

(4.8) $\dfrac{\partial P}{\partial t} = D_P \Delta P + k_1 TC - k_2 PS + k_3 P^*,$

(4.9) $\dfrac{\partial P^*}{\partial t} = k_2 PS - k_3 P^*,$

(4.10) $\dfrac{\partial S}{\partial t} = -k_4 S^{-1/2} f_{\text{dev}}(R, P, S, P^*) E(x) - k_2 PS + k_3 P^*,$

where $E(x)$ is a given exposure function; $E(x) \geq 0$. D_R, D_T, D_P, and the k_i are all positive constants. $E(x)$ is larger at the lighter portions of the film and smaller at the darker portions; at totally dark points $E(x) = 0$. The factor $S^{-1/2}$ in the last equation comes from the fact that $dV = k_5 S^{1/2} dS$ where dV is the volume element.

Boundary conditions for the diffusing species R, T, P are specified at each edge of the film, and flux equality is assumed at the gelatin-emulsion interfaces.

To complete the description of the system we must specify the function f_{dev}.

Observe that this model includes diffusion for R, T, and P; all other concentrations undergo chemical processes but not dynamical motion. The deep part of the chemistry is embodied in the term f_{dev}, which has yet to be fully quantified.

4.3. The bulk reaction problem

The solution of (4.4)–(4.10) with appropriate initial and boundary conditions is a formidable task; more so because f_{dev} is not quite known and must be "fitted" to experimental results. However, we can get some flavor of the analysis by concentrating on a special case, which deals only with the *bulk reaction*. In this model we ignore the silver halide grains. We simply assume that a substance A (the oxidized developer) diffuses through the emulsion and reacts only with another substance B distributed in the oil droplets. We take for simplicity $f_{\text{dev}} \equiv \gamma$, a positive constant. Then the equations

(4.11) $\dfrac{\partial A}{\partial t} = D \Delta A - kAB + \gamma E(x),$

(4.12) $\dfrac{\partial B}{\partial t} = -kAB$

hold in the solution. Obviously we want k positive. Here (4.11) and (4.12) correspond to (4.5) and (4.6) with A replacing T and B replacing C. For simplicity we consider the one-dimensional case first. Then (4.11) becomes

(4.13) $$\frac{\partial A}{\partial t} = D\frac{\partial^2 A}{\partial x^2} - kAB + \gamma E(x) \quad \text{if } 0 < x < L, \, t > 0.$$

We must give initial conditions

(4.14) $A(x,0) = A_0(x), \qquad 0 \le x \le L \quad (A_0 \ge 0),$

(4.15) $B(x,0) = B_0(x), \qquad 0 \le x \le L \quad (B_0 \ge 0)$

and boundary conditions, say,

(4.16) $A(0,t) = 0, \qquad A(L,t) = 0 \quad \text{for } t > 0.$

We wish to study the problem (4.12)–(4.16).

4.4. Analysis of the solution

Let R be the rectangle $0 < x < L$, $0 < t \le T$ (including its "top"), while C is the part of the boundary of R contained in the three lines $x = 0$, $x = L$ and $t = 0$, as illustrated in Fig. 4.4.

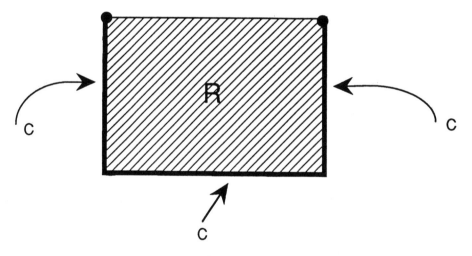

FIG. 4.4.

THEOREM 4.4.1. (The maximum principle). *If $A = A(x,t)$ is continuous in $R \cup C$, and satisfies the inequality*

(4.17) $$\frac{\partial A}{\partial t} - D\frac{\partial^2 A}{\partial x^2} + cA \ge 0 \quad in \ R,$$

where D is a positive function and c is any bounded function, and if

(4.18) $A \ge 0 \quad on \ C,$

then

(4.19) $$A \geq 0 \quad in \ R.$$

Proof. First consider the case where $c \geq 0$ and (4.17) is replaced by the *strict* inequality ">." Let us recall the fact that a continuous function on a closed bounded set must attain a maximum and a minimum value somewhere on that set. Repeating the ideas begun in Chapter 3, if the conclusion of the theorem were false, then there would be a point (x_0, t_0) in R where $A(x_0, t_0) < 0$. Hence the minimum of A would have to be attained at a point belonging to R, not C. At such a point $A < 0$, and hence

$$\frac{\partial A}{\partial t} \leq 0, \quad -\frac{\partial^2 A}{\partial x^2} \leq 0, \quad cA \leq 0,$$

thus forcing the left-hand side of (4.17) to be ≤ 0, contradicting the "strict" inequality (4.17). Since this is impossible, we must have $A \geq 0$ on all of $R \cup C$.

Next, we wish to remove the assumption that (4.17) is a *strict* inequality. We do this by introducing a "modified" function A. Let

$$u = u(x, t) = A + \varepsilon t,$$

where ε is a positive number, which will eventually be allowed to approach zero. Then

$$\frac{\partial u}{\partial t} - D\frac{\partial^2 u}{\partial x^2} + cu \geq \varepsilon + \varepsilon ct > 0 \quad in \ R.$$

Since $u \geq A \geq 0$ on C, we have, by what has just been proved, $u \geq 0$ in R, or $A \geq -\varepsilon t \geq -\varepsilon T$ in R. Now since A is *independent* of ε, we can let $\varepsilon \searrow 0$ and conclude again that $A \geq 0$ in R also.

To prove the assertion (4.19) for general c, we simply work with the function $\widetilde{A} = e^{-\alpha t}A$ which satisfies

$$\frac{\partial \widetilde{A}}{\partial t} - D\frac{\partial^2 \widetilde{A}}{\partial x^2} + (c + \alpha)\widetilde{A} \geq 0$$

and take $\alpha > \max\{-c\}$.

The maximum principle yields the following useful result.

THEOREM 4.4.2. (Comparison theorem). *If $u(x, t)$ and $v(x, t)$ are such that the function $A = u - v$ satisfies (4.17), (4.18), then $u \geq v$ in $0 \leq x \leq L$, $0 \leq t \leq T$.*

Note that if A satisfies the *strict* inequality

$$\frac{\partial A}{\partial t} - D\frac{\partial^2 A}{\partial x^2} < 0 \quad in \ R$$

(with $D > 0$), then A cannot have a local maximum at any *interior* point of R. For, at such a point

$$\frac{\partial^2 A}{\partial x^2} \leq 0, \qquad \frac{\partial A}{\partial t} = 0,$$

contradicting the strict inequality. Moreover, a local maximum cannot occur on the open interval

$$0 < x < L, \qquad t = T,$$

because at such a maximum point

$$\frac{\partial A}{\partial t} \geq 0, \qquad \frac{\partial^2 A}{\partial x^2} \leq 0,$$

again giving a contradiction. Thus no maximum can be attained in R. It is a very useful fact that the result just mentioned holds even under the following weak inequality.

THEOREM 4.4.3. (The strong maximum principle). *If $D > 0$ in R, $c(x,t)$ is a bounded nonnegative function in R and*

$$\frac{\partial A}{\partial t} - D\frac{\partial^2 A}{\partial x^2} + cA \leq 0 \quad in \ R,$$

then A cannot attain a nonnegative maximum in R, unless A is identically constant.

(This will not be proved here, but a version for equations *independent of t* will be proved in the appendix to this chapter.)

Next, consider the problem (4.12)–(4.16). We can write

$$(4.20) \qquad B(x,t) = B_0(x)\exp\left(-k\int_0^t A(x,s)ds\right).$$

Equation (4.13) implies the differential inequality (4.17) with $c = kB$. Hence, by the comparison theorem,

$$A \geq 0.$$

It follows from (4.12) that $\partial B/\partial t \leq 0$, that is,

$$B(x,t) \quad \text{is monotone nonincreasing in } t.$$

PROBLEM 4.4.1. Solve the system (4.12), (4.13) numerically, with initial conditions (4.14) and (4.15) and boundary conditions (4.16), for a time interval of two to three *minutes*. Try the following numerical values:

$$D = 100 \text{microns}^2/\text{sec}, \qquad (\text{microns} = \mu m)$$

$$\gamma = 7.5 \times 10^{-12} \text{moles/micron sec},$$

$$k = 6.6 \times 10^{12} \text{microns/moles sec},$$

$$L = 15 \text{ centimeters} = 1.5 \times 10^5 \mu m,$$

$$E = 1 \text{ for} \begin{cases} 1.0 \times 10^4 \text{ microns} < x < 2.0 \times 10^4 \text{ microns} \\ \text{or } 3 \times 10^4 \text{ microns} < x < 4.0 \times 10^4 \text{ microns}, \end{cases}$$

$$E = 0 \quad \text{otherwise},$$

$$A_0(x) \equiv 0,$$

$$B_0(x) = 1.125 \times 10^{-11} \text{moles/micron}.$$

Recall that A (previously T) represents the oxidized developer, while B (previously C) represents the coupler. What we really want to study is the dye Y represented by equation (4.7). If Y is taken to be initially zero, how can we compute its concentration after two to three minutes? Note that because of (4.6) and (4.7) the rate of change of the dye (Y) plus coupler ($B = C$) must be zero. Hence

(4.21) $$Y = B_0 - B.$$

4.5. Late development of film

Problem 4.4.1 is a numerical study of the development process of the film, lasting two to three minutes. What happens if the person developing the film is absentminded and leaves the film in the solution (say overnight)? To answer that question we should study the long-term behavior with the quantities A, B, Y, as $t \to \infty$. Recall that $B(x,t)$ is monotone nonincreasing as t increases.

Denote by $\widetilde{B}(x)$ the limit of $B(x,t)$ as $t \to \infty$. (Why does this limit exist?)

We can also show that as $t \to \infty$ the function $A(x,t)$ converges to some function $\widetilde{A}(x)$, and that $\partial A(x,t)/\partial t$ converges to zero. It then follows from (4.13) that

(4.22) $$- D\frac{\partial^2 \widetilde{A}}{\partial x^2} + k\widetilde{A}\widetilde{B} = \gamma E(x).$$

Assume that $E(x)$ is not identically equal to zero.

From the strong maximum principle for (4.22) (see the Appendix to this chapter) we then deduce $\widetilde{A}(x) > 0$ for $0 < x < L$. (Can you deduce this directly?) Therefore, if we let $t \to \infty$ in (4.20) we conclude that

$$\widetilde{B}(x) = \lim_{t \to \infty} B(x,t) = 0.$$

Thus we arrive at the final form of the equation for $\widetilde{A}$:

$$(4.23) \qquad\qquad\qquad -D\frac{\partial^2 \widetilde{A}}{\partial x^2} = \gamma E(x)$$

in $0 < x < L$ with
$$(4.24) \qquad\qquad\qquad \widetilde{A}(0) = 0, \qquad \widetilde{A}(L) = 0.$$

We could study the behavior of $\widetilde{A}(x)$ by means of (4.23), (4.24). However, what we are really interested in is the long-term behavior of the dye concentration Y. Using (4.21) and the fact that $\widetilde{B}(x) = \lim_{t\to\infty} B(x,t) = 0$, we see that

$$\widetilde{Y}(x) = \lim_{t\to 0} Y(x,t) = B_0(x).$$

Thus the distribution of the dye, which is what we see in a picture, is independent of the exposure $E(x)$ in that situation. In particular, if the coupler C is uniformly distributed initially (as in the case in the numerical example) the developed picture will be a uniform grey (or other color)! This is exactly what we would expect to happen if we forget the film in the development solution overnight!

   Note. The astute reader will note that if the exposure $E(x)$ is discontinuous (as in Problem 4.4.1) then $\widetilde{A}$ need not have two continuous derivatives, and the strong maximum principle as stated in the Appendix cannot be applied directly.

   PROBLEM 4.5.1. Show that in the case where $E(x)$ has only jump discontinuities (and is nonnegative) then the maximum principle still holds.
   Hint. Use the fact that the solution will have a continuous first derivative, together with Theorem 4.9.2 of the Appendix to this chapter.

4.6. Implicit methods for solving system (4.12)–(4.16) numerically

Let us begin by considering the problem of numerically solving the following simple initial value problem:

$$\begin{cases} \dfrac{\partial}{\partial t} A = D\dfrac{\partial^2}{\partial x^2} A, & x \in \mathbb{R}, \quad 0 < t < T, \\[2mm] A(x,0) = A_0(x), & x \in \mathbb{R}. \end{cases}$$

Perhaps the simplest numerical scheme for the above problem is obtained by using a centered approximation to $\partial^2/\partial x^2\, A$ and a *forward* approximation to $\partial/\partial t\, A$, namely,

$$\frac{\partial^2}{\partial x^2}\, A(x = m\Delta x) \approx \frac{1}{(\Delta x)^2}\Big(A((m-1)\Delta x) - 2A(m\Delta x) + A((m+1)\Delta x)\Big),$$

$$\frac{\partial}{\partial t}\, A(t = n\Delta t) \approx \frac{1}{\Delta t}\Big(A((n+1)\Delta t) - A(n\Delta t)\Big).$$

We obtain the following scheme ($\mathbb{Z}$ = set of all integers):

$$\begin{cases} \dfrac{A_m^{n+1} - A_m^n}{\Delta t} = D \dfrac{A_{m-1}^n - 2A_m^n + A_{m+1}^n}{(\Delta x)^2}, & m \in \mathbb{Z}, \quad n \geq 0, \\ A_m^0 = A_0(m\Delta x), & m \in \mathbb{Z}. \end{cases}$$

If we set $\mu = D\,\Delta t/\Delta x^2$, we can rewrite the above scheme as follows:

$$\begin{cases} A_m^{n+1} = \mu A_{m-1}^n + (1 - 2\mu) A_m^n + \mu A_{m+1}^n, & m \in \mathbb{Z}, \quad n \geq 0, \\ A_m^0 = A_0(m\Delta x), & m \in \mathbb{Z}. \end{cases}$$

Let us see if this scheme satisfies the von Neumann stability criterion. Proceeding as in §2.7, we obtain

$$\xi = 1 - 4\mu \sin^2 \beta/2.$$

Thus, $|\xi| \leq 1$ if and only if $\mu \leq \frac{1}{2}$, that is, if and only if

$$\Delta t \leq \frac{1}{2D} \cdot (\Delta x)^2.$$

This restriction on the size of the time step is not desirable because it forces us to multiply by four the number of time steps each time we double the number of gridpoints in space. To overcome this difficulty, we consider the following scheme:

$$\begin{cases} \dfrac{A_m^n - A_m^{n-1}}{\Delta t} = D \dfrac{A_{m-1}^n - 2A_m^n + A_{m+1}^n}{(\Delta x)^2}, & m \in \mathbb{Z} \quad n \geq 1, \\ A_m^0 = A_0(m\Delta x), & m \in \mathbb{Z} \end{cases}$$

which is obtained by using the same centered approximation to $\partial^2/\partial x^2\, A$ and a *backward* approximation to $\partial/\partial t\, A$, namely,

$$\frac{\partial}{\partial t}\, A(t = n\Delta t) = \frac{1}{\Delta t}(A(n\Delta t) - A((n - 1)\Delta t)).$$

We can verify that the above scheme always satisfies the von Neumann criterion of stability. Indeed in this case, we have

$$\xi = \frac{1}{1 + 4\mu \sin^2 \beta/2} \leq 1 \quad \forall \; |\beta| \leq \pi.$$

Thus, this scheme is *unconditionally stable*.

The price we must pay for this property is that now we cannot compute A_m^{n+1} *explicitly* from the values $\{A_m^n\}_{m \in \mathbb{Z}}$ as with the first scheme. In the present case, the values $\{A_m^{n+1}\}_{m\in\mathbb{Z}}$ are defined *implicitly* in terms of the values $\{A_m^n\}_{m\in\mathbb{Z}}$ as follows:

$$\begin{cases} -\mu A_{m-1}^{n+1} + (1+2\mu)A_m^{n+1} - \mu A_{m+1}^{n+1} = A_m^n, & m \in \mathbb{Z}, \quad n \geq 0, \\ A_m^0 = A_0(m\Delta x), & m \in \mathbb{Z}. \end{cases}$$

Let us now discretize the following initial boundary value problem:

$$\frac{\partial}{\partial t} A = D\frac{\partial^2 A}{\partial x^2}, \qquad 0 < x < L, \quad 0 < t < T,$$

$$A(x,0) = A_0(x), \qquad 0 < x < L,$$

$$A(0,t) = A(L,t) = 0, \qquad 0 \leq t \leq T$$

by using our implicit scheme. We obtain

$$\begin{cases} -\mu A_{m-1}^{n+1} + (1+2\mu)A_m^{n+1} - \mu A_{m+1}^{n+1} = A_m^n & \left\{\begin{array}{l} m = 1,\ldots,M, \\ n = 0,\ldots,N, \end{array}\right\}, \\ A_0^{n+1} = A_{M+1}^{n+1} = 0 \\ A_m^0 = A_0(m\Delta x), & m = 0,\ldots,M+1. \end{cases}$$

This scheme can be written in matrix form as follows:

$$MA^{n+1} = A^n,$$

where

$$M = \begin{bmatrix} (1+2\mu) & -\mu & 0 & \ldots & 0 & 0 \\ -\mu & (1+2\mu) & -\mu & \ldots & 0 & 0 \\ 0 & -\mu & (1+2\mu) & \ldots & 0 & 0 \\ \ldots & \ldots & \ldots & \ldots & \ldots & \ldots \\ 0 & 0 & 0 & \ldots & (1+2\mu) & -\mu \\ 0 & 0 & 0 & \ldots & -\mu & (1+2\mu) \end{bmatrix},$$

$$A^{n+1} = \begin{bmatrix} A_1^{n+1} \\ A_2^{n+1} \\ A_3^{n+1} \\ \vdots \\ A_{M-1}^{n+1} \\ A_M^{n+1} \end{bmatrix}, \qquad A^n = \begin{bmatrix} A_1^n \\ A_2^n \\ A_3^n \\ \vdots \\ A_{M-1}^n \\ A_M^n \end{bmatrix}, \qquad n \geq 0$$

$$A_m^0 = A_0(m\Delta x), \qquad m = 0, \ldots, M+1.$$

To numerically solve this equation, we first find the so-called "LU-decomposition" of M (L is a lower triangular matrix and U is an upper triangular matrix), $M = LU$, namely,

$$
\begin{bmatrix}
(1+2\mu) & -\mu & 0 & \ldots & 0 \\
-\mu & (1+2\mu) & -\mu & \ldots & 0 \\
0 & -\mu & (1+2\mu) & \ldots & 0 \\
\ldots & \ldots & \ldots & \ldots & \ldots \\
0 & 0 & 0 & \ldots & (1+2\mu)
\end{bmatrix}
$$

$$
=
\begin{bmatrix}
1 & 0 & 0 & \ldots & 0 \\
\ell_{21} & 1 & 0 & \ldots & 0 \\
0 & \ell_{32} & 1 & \ldots & 0 \\
\ldots & \ldots & \ldots & \ldots & \ldots \\
0 & 0 & 0 & \ldots & 1
\end{bmatrix}
\begin{bmatrix}
u_{11} & u_{12} & 0 & \ldots & 0 \\
0 & u_{22} & u_{23} & \ldots & 0 \\
0 & 0 & u_{33} & \ldots & 0 \\
\ldots & \ldots & \ldots & \ldots & \ldots \\
0 & 0 & 0 & \ldots & u_{MM}
\end{bmatrix}.
$$

The quantities ℓ_{ij} and u_{ij} are very easily computed recursively (first row of M):

$$
\begin{cases}
(1+2\mu) = u_{11}, \\
-\mu = u_{12}
\end{cases}
$$

(second row of M):

$$
\begin{cases}
-\mu = \ell_{21}u_{11} & \Rightarrow \ell_{21} = -\mu/u_{11} \\
(1+2\mu) = \ell_{21}u_{12} + u_{22} & \Rightarrow \quad u_{22} = (1+2\mu) - \ell_{21}u_{12} \\
-\mu = u_{23}
\end{cases}
$$

and so on.

Once the above computations are done, to solve the matrix equation

$$MA^{n+1} = LUA^{n+1} = A^n,$$

we first solve for Y (note that $UA^{n+1} = Y$!)

$$LY = A^n,$$

and then we solve for A^{n+1}:

$$UA^{n+1} = Y.$$

With the above ideas, we can now devise a numerical scheme to solve the system (4.12)–(4.16). For example, we can use the implicit method of discretization

$$\frac{B_m^{n+1} - B_m^n}{\Delta t} = -kA_m^n B_m^{n+1}, \qquad m = 1, \dots, M, \quad n \geq 0,$$

$$\frac{A_m^{n+1} - A_m^n}{\Delta t} = D\frac{A_{m-1}^{n+1} - 2A_m^{n+1} + A_{m+1}^{n+1}}{\Delta x^2} - kA_m^n B_m^{n+1} + \gamma E_m,$$

$$m = 1, \dots, M, \quad n \geq 0,$$

$$A_0^{n+1} = A_{M+1}^{n+1} = 0, \qquad n \geq 0,$$

$$A_m^0 = A_0(m\Delta x), \qquad m = 0, \dots, M+1,$$

$$B_m^0 = B_0(m\Delta x), \qquad m = 0, \dots, M+1.$$

PROBLEM 4.6.1. Solve Problem 4.4.1 by using this implicit method.

PROBLEM 4.6.2. Implicit Methods for Diffusion Equations. Consider the problem of finding the solution A of

$$A_t - A_{xx} = \sin(\pi x) \quad \text{in } (0, T) \times (0, 1),$$
$$A(0, t) = A(1, t) = 0 \quad \text{for } t \in [0, T],$$
$$A(x, t = 0) = 0 \quad \text{for } x \in [0, 1].$$

(a) Obtain the solution A.
Hint. The solution A is of the form $f(t)\sin(\pi x)$.

(b) Discretize the problem under consideration by using the backward Euler method in time and the centered difference method in space. Write the resulting equations in matrix form as $M\mathbb{A} = F$, where $\mathbb{A}$ is the matrix of unknowns at time $t = n\Delta t$. Write a routine to obtain the LU decomposition of M and another to solve the equation $LU\mathbb{A} = F$.

(c) Use the above routines to obtain a numerical approximation to A at time $T = 10$. Begin by taking $\Delta t = 10$ and $\Delta x = 1/10$. Compute the maximum error between the exact solution and your approximate solution at the grid points. Also, save the CPU time your program used to perform the computation. Now, for the same Δt, repeat the exercise with $\Delta x = 1/20$, $\Delta x = 1/40, \dots$ until the error does not diminish anymore. Next take $\Delta t = 5$ and repeat the above calculations. Then, continue with $\Delta t = 5/2$, $\Delta t = 5/4, \dots$ until the error you obtain is less than or equal to 10^{-5}. Display your results in a table.

(d) Discretize the problem under consideration by using the forward Euler method in time and the centered difference method in space. Pick $\Delta x = 1/10$, compute the maximum Δt for which the stability holds. Compute the approximate solution; get the maximum error and the CPU time. Now take $\Delta x = 1/20$, $\Delta x = 1/40, \dots$ and repeat the exercise. Display your results in a table.

(e) By comparing the tables obtained in the two previous exercises, what can we say about the performance of the numerical schemes?

4.7. Summary

We began by describing the physical process of color photography, and introducing a complicated model given by (4.4)–(4.10). This model was simplified considerably in §4.3. Properties of this system were studied by means of the maximum principle, in §4.4, and a numerical study of the development process was given as a problem. The limiting behavior as $t \to \infty$ was discussed. Finally, implicit methods for numerical solution of the simplified model were developed.

4.8. Appendix: Proof of the strong maximum principle[4]

A function $u(x)$ that is continuous on the closed interval $[a, b]$ takes on its maximum at a point on this interval. If $u(x)$ has a continuous second derivative, and if u has a relative nonnegative maximum at some point c between a and b, then we know from elementary calculus that

$$(4.25) \qquad u(c) \geq 0, \ u'(c) = 0, \quad \text{and} \quad u''(c) \leq 0.$$

Suppose that in an open interval (a, b), u is known to satisfy a differential inequality of the form

$$(4.26) \qquad L[u] \equiv u'' + g(x)u' - h(x)u > 0,$$

where $g(x)$ is any bounded function and $h(x)$ is any nonnegative function. Then it is clear that relations (4.25) cannot be satisfied at any point c in (a, b). Consequently, whenever (4.26) holds, the nonnegative maximum of u in the interval cannot be attained anywhere except at the endpoints a or b. We have here the simplest case of a *strong maximum principle*.

THEOREM 4.8.1. (One-dimensional strong maximum principle). *Suppose $u = u(x)$ satisfies the differential inequality*

$$(4.27) \qquad L[u] \equiv u'' + g(x)u' - h(x)u \geq 0 \quad \text{for } a < x < b,$$

with $g(x)$ and $h(x)$ bounded functions and $h \geq 0$. If $u(x) \leq M$ in (a, b) and if the maximum M of u is nonnegative and is attained at an interior point c of (a, b), then $u \equiv M$.

Proof. We suppose that $u(c) = M$ and that there is a point d in (a, b) such that $u(d) < M$. We will show that this leads to a contradiction. First consider the case $d > c$. We define the function

$$z(x) = e^{\alpha(x-c)} - 1$$

---

[4]We follow the treatment in M. F. Protter and H. F. Weinberger, *Maximum Principles of Differential Equations*, Prentice Hall, 1967.

with α a positive constant to be determined. Note that $z(x) < 0$ for $a < x < c$, that $z(x) > 0$ for $c < x < b$, and that $z(c) = 0$. (See Fig. 4.5.)

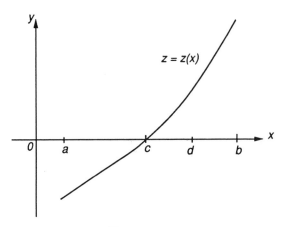

FIG. 4.5.

A simple computation yields

$$L[z] \equiv z'' + g(x)z' - h(x)z = \{\alpha[\alpha + g(x)] - h(x)\}e^{\alpha(x-c)} + h(x).$$

We choose α so large that $L[z] > 0$ for $a < x < d$. That is, we select α so that it satisfies the inequality

$$\alpha > -g(x) + \frac{h(x)}{\alpha};$$

we can always do this since $g(x)$ and $h(x)$ are bounded. We now define

$$w(x) = u(x) + \varepsilon z(x),$$

where ε is a positive constant chosen so that it satisfies the inequality

$$\varepsilon < \frac{M - u(d)}{z(d)}.$$

The assumption $u(d) < M$ and the fact that $z(d) > 0$ make it possible to find such an ε. Then, since z is negative for $a < x < c$, we have

$$w(x) < M \quad \text{for} \ a < x < c;$$

by the definition of ε,

$$w(d) = u(d) + \varepsilon z(d),$$

$$< u(d) + M - u(d),$$

so that
$$w(d) < M.$$

At the point c,
$$w(c) = u(c) + \varepsilon z(c) = M.$$

Hence w has a maximum greater than or equal to M that is attained at an interior point of the interval (a, d). However,

$$L[w] = L[u] + \varepsilon L[z] > 0,$$

so that by our previous result concerning the strict inequality (4.26), w cannot attain its maximum in (a, d). We thereby reach a contradiction.

If $d < c$, we use the auxiliary function

$$z = e^{-\alpha(x-c)} - 1$$

with $\alpha > g(x) + h(x)/\alpha$ to reach the same conclusion.

By applying Theorem 4.9.1 to $(-u)$ we have the *strong minimum principle*, which asserts that a nonconstant function satisfying the differential inequality $L[u] \leq 0$ cannot attain its nonpositive minimum at an interior point.

THEOREM 4.8.2. *Suppose u is a nonconstant function that satisfies the inequality $u'' + g(x)u' - h(x)u \geq 0$ in (a, b) and has one-sided derivatives at a and b, and suppose g and h are bounded on the closed interval $[a, b]$ and $h \geq 0$. If the maximum of u occurs at $x = a$ and is nonnegative, then $u'(a) < 0$. If the maximum occurs at $x = b$ and is nonnegative, then $u'(b) > 0$.*

Proof. Suppose that $u(a) = M$, that $u(x) \leq M$ for $a \leq x \leq b$, and that for some point d in (a, b) we have $u(d) < M$. Once again we define an auxiliary function

$$z(x) = e^{\alpha(x-a)} - 1 \quad \text{with } \alpha > 0.$$

We select $\alpha > -g(x) + h(x)/\alpha$ for $a \leq x \leq d$ so that $L[z] > 0$. Next, we form the function

$$w(x) = u(x) + \varepsilon z(x)$$

with ε chosen so that

$$0 < \varepsilon < \frac{M - u(d)}{z(d)}.$$

Because $L[w] > 0$, the maximum of w in the interval $[a, d]$ must occur at one of the ends. We have

$$w(a) = M > w(d),$$

so that the maximum occurs at a. Therefore, the one-sided derivative of w at a cannot be positive:

$$w'(a) = u'(a) + \varepsilon z'(a) \leq 0.$$

However,
$$z'(a) = \alpha > 0,$$
and therefore
$$u'(a) < 0,$$
which is the desired result.

If the maximum occurs at $x = b$, the argument is similar.

References

The model (4.4)–(4.10) was developed by David Ross and presented in

[1] A. Friedman, *Mathematics in Industrial Problems, Part* 2, Chap. 10, IMA Vol. Math. Appl., Vol. 24, Springer-Verlag, New York, 1989.

The model is based on

[2] R. Matejec and R. Mayer, Zeit. fur Wissen, Photographic, 57 (1963), pp. 18–47.

Chapter 5

How Does a Catalytic Converter Function?

5.1. Background

A catalytic converter is a device located in the exhaust system of the automobile, approximately underneath the driver seat, between the engine output and the exhaust tailpipe. As pollutant gases flow out of the engine, they pass through the catalytic converter and undergo chemical processes by which they are converted into relatively harmless gases.

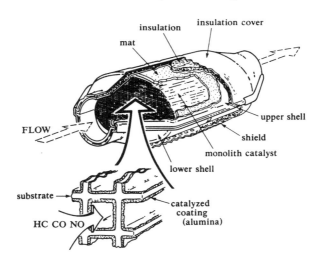

FIG. 5.1. *Ceramic monolith catalytic converter.*[1]

Because of new government regulations for emission control contained in revisions to the Clean Air Act, there is a renewed urgency to improve the performance of catalytic converters. The goal is to predict the transient tailpipe

[1]A. Friedman, *Mathematics in Industrial Problems*, IMA Vol. Math. Appl., Vol. 38, Springer-Verlag, New York, 1991, Fig. 7.1, p. 71. ©1991 Springer-Verlag. Reprinted with permission.

emission, or converter output, given the engine output for any driving cycle of interest. The main cycles of interest from the standpoint of emission and thermal stress are the warm-up (which causes the largest amount of pollution), sustained heavy load, and engine misfiring.

The present day converter is usually a ceramic monolith (i.e., a piece of ceramic). It is a tubular reactor, as shown in Fig. 5.1. Gas flows through the passes and reacts on the surface of the tubular walls.

5.2. The model

The catalytic converter to be studied here is based on oxidation reaction of CO, hydrocarbons and H_2; NO is not included in the reaction. The oxidation reactions are

$$CO + \tfrac{1}{2} O_2 \rightarrow CO_2,$$

(5.1)
$$C_3H_6 + \tfrac{9}{2} O_2 \rightarrow 3CO_2 + 3H_2O,$$
$$H_2 + \tfrac{1}{2} O_2 \rightarrow H_2O.$$

Let $i = 1$ stand for CO, $i = 2$ for C_3H_6, $i = 3$ for H_2, and $i = 4$ for O_2. Propylene, C_3H_6, represents hydrocarbon species whose reactions are fast enough to be important in the modeling.

The specific reaction rates R_i for the oxidation of CO, C_3H_6, and H_2 have the form
(5.2) $$R_i = k_i c_i c_4 / G \qquad (i = 1, 2, 3),$$

where c_i is the concentration of the ith species and G is given by the experimental formula

(5.3) $$G = T(1 + K_1 c_1 + K_2 c_2)^2 \cdot (1 + K_3 c_1^2 c_2^2)(1 + K_4 c_{NO}^{0.7}),$$

where c_{NO} is the concentration of NO and k_i, K_j depend only on the temperature T; k_1, k_2, K_4 are monotone increasing in T and K_1, K_2, K_3 are monotone decreasing in T.

Most catalytic converters are designed to operate at the stoichiometrically balanced air/fuel ratio for the simultaneous conversion of carbon monoxide, hydrocarbons, and nitrogen oxides in automobile exhaust; this means that there is conservation of mass and energy, which, in view of (5.1), takes the form
(5.4) $$R_4 = 0.5R_1 + 4.5R_2 + 0.5R_3.$$

Let

$$T_g = \text{temperature of the gas},$$

$$T_s = \text{temperature of the solid,}$$

$$c_{gi} = \text{concentration of species } i \text{ in the bulk gas stream,}$$

$$c_{si} = \text{concentration of species } i \text{ on the solid surface.}$$

If we ignore channel-to-channel variations within the converter and denote by x the length parameter along the parallel tubes, then the following system of differential equations holds:

$$(5.5) \qquad C(T_s)\frac{\partial T_s}{\partial t} = \lambda\frac{\partial^2 T_s}{\partial x^2} + h(T_g)(T_g - T_s) + a\sum_{j=1}^{3} R_j(T_s, \vec{c}_s),$$

$$(5.6) \qquad w\frac{\partial T_g}{\partial x} = h(T_g)(T_s - T_g),$$

$$(5.7) \qquad -w\frac{\partial c_{gi}}{\partial x} = K_{mi}(T_g)(c_{gi} - c_{si}) \qquad (i = 1, \ldots, 4),$$

for $0 < x < \infty$, $t > 0$, where $x = 0$ is the initial point of the converter; we assume that the converter is very long, i.e., it extends to the entire interval $0 < x < \infty$.

The chemistry between the gases on the solid surface and the bulk gas flow is expressed by the algebraic equations

$$(5.8) \qquad K_{mi}(T_g)(c_{gi} - c_{si}) = aR_i(T_s, \vec{c}_s) \qquad (i = 1, \ldots, 4).$$

Here a is the catalyst activity factor, R_j are the specific reaction rates defined in (5.2)–(5.4), h is the heat transfer coefficient, $C(T_s)$ is the specific heat of the solid, K_{mi} is the mass transfer coefficient for species i, and w is the mass flow rate; because of the conservation of mass, w does not depend on x, i.e., $w = w(t)$. The function $w(t)$ is known. We are also given the boundary values of

$$(5.9) \qquad\qquad T_s, T_g, \text{ and } \vec{c}_g \text{ at } x = 0, \qquad t > 0,$$

and the initial data

$$(5.10) \qquad\qquad T_s \text{ at } t = 0 \quad \text{for all } x;$$

T_s at $t = 0$, the ambient temperature, is usually a constant.

The system (5.5)–(5.10) has many common features with the system of reaction-diffusion equations that appeared in Chapter 4 in modeling the color film negative development. Note that the temperature T_s in the solid is subject to diffusion (as in the usual case of the heat equation). However, in the gas bulk flow the flow velocity is so large that the diffusion of the gas temperature can be ignored.

5.3. The control problem

Catalytic converters are designed with the goal of minimizing the concentration of the polluting gases, as measured at a preassigned location, say $x = L$, along the device. Let us consider this problem in the most important case of the warm-up cycle. You start your car when the engine is cold, say at temperature $T_s = 300°$ (in K). At such low temperature the reaction rates are very slow, and consequently, the converter is not performing efficiently in converting the polluting gases into the relatively harmless ones. To speed up the conversion rate we may artificially raise the temperature at the endpoint $x = 0$, say to

$$T_s|_{x=0} = 300 + S(t); \qquad S(t) \geq 0.$$

The function $S(t)$ is called a *control function*.

For mechanical reasons we must impose some restrictions on $S(t)$; for instance,

$$(5.11) \qquad \int_0^{t_0} S(t)dt \leq M, \qquad S(t) \leq N \quad \text{for } 0 \leq t \leq t_0$$

if t_0 is the duration of the warm-up interval; M and N are given positive constants.

Our goal is to reduce concentration c_{gi} at $x = L$. Since each gas species counts differently in a test for pollution control, what we want to minimize is an expression of the form

$$(5.12) \qquad \sum_{j=1}^{3} \lambda_j \int_0^{t_0} c_{gj}(L, t)dt,$$

where the λ_j are some specified positive constants.

For any choice of the control S, we must solve the system (5.5)–(5.10) and then compute the expression in (5.12). We denote this expression by $J(S)$, i.e.,

$$(5.13) \qquad J(S) = \sum_{j=1}^{3} \lambda_j \int_0^{t_0} c_{gj}(L, t)dt;$$

we call it the *cost function*.

Let us also introduce the set $\mathcal{A}$ of control functions

$$(5.14) \qquad \mathcal{A} = \left\{ S(t), \ 0 \leq t \leq t_0; \ S(t) \text{ is piecewise continuous and } 0 \leq S(t) \leq N \right.$$
$$\left. \text{and } t \int_0^{t_0} S(t)dt \leq M \right\}.$$

Then we are interested in solving the following optimal control problem:

$$(5.15) \qquad \underset{S \in \mathcal{A}}{\text{minimize}} J(S).$$

5.4. A simplified model

The control problem for the model (5.5)–(5.10) involves very intensive computations. What we want to do here is consider a simpler version of (5.5)–(5.10) that still maintains the important aspects of the full model and analyze it mathematically.

We assume that $h(T_g) = 0$ in (5.6), and that there is only one species, with concentration c; $c = c_s$ on the solid and $c = c_g$ in the gas bulk. We take the reaction rate to be

$$R = R(T_s, c_s) = (T_s - 300)c_s.$$

Thus, at the cold engine temperature, the reaction rate is zero, and the reaction rate increases with the temperature. Assuming also that

$$\lambda = 1, \quad a = 1, \quad C(T_s) \equiv 1, \quad K_{m_i} \equiv 1, \quad w = 1$$

and setting

$$T = T_s - 300, \qquad u = c_g,$$

the system (5.5)–(5.8) reduces to

$$(5.16) \qquad \frac{\partial T}{\partial t} = \frac{\partial^2 T}{\partial x^2} + T c_s,$$

$$(5.17) \qquad -\frac{\partial u}{\partial x} = u - c_s,$$

$$(5.18) \qquad u - c_s = T c_s.$$

From the last equation we obtain

$$c_s = \frac{u}{1 + T}.$$

Substituting this into (5.16) and (5.17) we obtain

$$(5.19) \qquad \frac{\partial T}{\partial t} = \frac{\partial^2 T}{\partial x^2} + \frac{T}{1 + T} u, \qquad 0 < x < \infty, \quad 0 < t < T,$$

$$(5.20) \qquad -\frac{\partial u}{\partial x} = \frac{T}{1 + T} u, \qquad 0 < x < \infty, \quad 0 < t < T.$$

We also have the initial conditions

$$(5.21) \qquad T(x, 0) = 0$$

and the boundary conditions

(5.22) $u(0,t) = u_0$ (u_0 positive constant),

(5.23) $T(0,t) = S(t)$,

where u_0 is the concentration of the gas as it enters the converter. We must impose a condition on $T(x,t)$ as $x \to \infty$, say: for any finite time interval $0 \le t \le t_0$,

(5.24) $T(x,t) \to 0$ as $x \to \infty$, uniformly in t for $0 \le t \le t_0$.

We wish to study the optimal control problem (5.15) for the system (5.19)–(5.24), where

(5.25) $$J(S) = \int_0^{t_0} u(L,t)dt.$$

First we need to develop some ideas from the calculus of variations and control theory.

5.5. The calculus of variations

In calculus one learns how to find maxima and minima of functions of one and several variables. One important tool is to find *necessary* conditions for *relative* or *local* maxima or minima (also called extrema). Thus to find the points where a local extremum of $f(x,y)$ *may* occur we set $\partial f/\partial x$ (x,y) and $\partial f/\partial y$ (x,y) equal to zero and solve for (x,y). In other words we look for points (x_0, y_0) at which the directional derivative of f in *all* directions is zero.

The calculus of variations tackles the problem of finding maxima and minima of functions of *functions* rather than functions of points. A function of functions is called a *functional*.

One of the first problems in the calculus of variations was a problem proposed by Johann Bernoulli near the end of the seventeenth century. Suppose two points are fixed in a vertical plane. What curve joining them will have the property that a particle sliding on the curve without friction from the higher point to the lower point under the influence of gravity will do so in the shortest possible time? This problem is called the *brachistochrone* problem from the Greek words meaning *shortest time*.

A mathematical formulation is obtained as follows. Let the particle P move along any curve given by the function $y = y(x)$. The speed v of the particle is given by

$$\frac{1}{2} v^2 = gy, \quad \text{i.e.,} \quad \frac{ds}{dt} = \sqrt{2g} \sqrt{y}.$$

Thus

$$dt = \frac{1}{\sqrt{2g}} \frac{ds}{\sqrt{y}} = \frac{1}{\sqrt{2g}} \sqrt{\frac{dx^2 + dy^2}{y}}$$

$$= \frac{1}{\sqrt{2g}} \sqrt{\frac{1 + (y')^2}{y}} \, dx.$$

See Fig. 5.2.

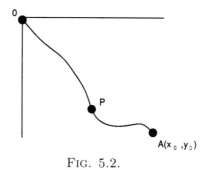

FIG. 5.2.

The total time required to reach the final point $A(x_0, y_0)$ from the initial point $(0,0)$ is

$$T = \frac{1}{\sqrt{2g}} \int_0^{x_0} \sqrt{\frac{1 + (y')^2}{y}} \, dx.$$

Thus T is a *functional* of the curve $y(x)$ connecting the two fixed points $(0,0)$ and A.

Setting

$$\frac{1}{\sqrt{2g}} \sqrt{\frac{1 + (y')^2}{y}} = f(y, y')$$

we may thus pose the brachistochrone problem as follows.

Among all the "smooth" functions $y(x)$ satisfying $y(0) = 0$ and $y(x_0) = y_0$, which are minimizers of the functional

$$J(y(x)) = \int_0^{x_0} f(y(x), y'(x)) dx?$$

5.6. The Euler–Lagrange equation

It turns out we can give a general and far-reaching answer to problems of this kind. Suppose we wish to determine which (if any) function $y = y(x)$

maximizes (or minimizes) the functional

$$I = \int_{x_1}^{x_2} f(x, y(x), y'(x)) dx$$

subject to the conditions

$$y(x_1) = y_1, \qquad y(x_2) = y_2.$$

Suppose there exists a function $y(x)$ that makes I an extremal. We are going to differentiate I *in the direction* of an (arbitrary) function $\eta(x)$. We take $\eta(x)$ to be differentiable and to satisfy the conditions $\eta(x_1) = 0$, $\eta(x_2) = 0$. We now consider the neighboring curve to $y(x)$

$$Y(x, \varepsilon) = y(x) + \varepsilon \eta(x).$$

We have

$$Y(x_1, \varepsilon) = y_1, \quad Y(x_2, \varepsilon) = y_2, \quad \text{and} \quad Y(x, 0) = y(x).$$

As we vary ε, we obtain a family of curves passing through the fixed points (x_1, y_1) and (x_2, y_2). For $\varepsilon = 0$ we obtain the original curve $y(x)$. We can think of the functional I as a function $I(\varepsilon)$ of ε:

$$I(\varepsilon) = \int_{x_1}^{x_2} f(x, y(x) + \varepsilon \eta(x), y'(x) + \varepsilon \eta'(x)) dx.$$

We are assuming that I is an extremal for $\varepsilon = 0$ (keeping the function $\eta(x)$ fixed for the moment). From ordinary calculus we therefore set

$$\frac{dI}{d\varepsilon}\bigg|_{\varepsilon=0} = 0.$$

Assuming the continuity of $\partial f / \partial y$ and $\partial f / \partial y'$, we obtain, by integration by parts,

$$I'(0) = \int_{x_1}^{x_2} \left(\frac{\partial f}{\partial y} \eta + \frac{\partial f}{\partial y'} \eta' \right) dx$$

$$= \int_{x_1}^{x_2} \frac{\partial f}{\partial y} \eta dx + \frac{\partial f}{\partial y'} \eta(x) \bigg|_{x=x_1}^{x=x_2} - \int_{x_1}^{x_2} \frac{d}{dx} \left(\frac{\partial f}{\partial y'} \right) \eta dx.$$

Since $\eta(x_1) = \eta(x_2) = 0$, we obtain

$$0 = I'(0) = \int_{x'}^{x_2} \left[\frac{\partial f}{\partial y} - \frac{d}{dx} \left(\frac{\partial f}{\partial y'} \right) \right] \eta(x) dx.$$

Now we vary $\eta(x)$. Since the quantity in square brackets is independent of the choice of $\eta(x)$, and $\eta(x)$ is arbitrary, we conclude that the bracketed expression must vanish identically. (Prove it!)

Thus a necessary condition for the function $y(x)$ to be an extremum of the functional

$$I = \int_{x_1}^{x_2} f(x, y(x), y'(x)) dx$$

(subject to $y(x_1) = y_1$, $y(x_2) = y_2$) is that $y(x)$ satisfy the so-called Euler *(or* Euler–Lagrange) *equation*

$$\frac{d}{dx} \left(\frac{\partial f}{\partial y'} \right) - \frac{\partial f}{\partial y} = 0.$$

PROBLEM 5.6.1. Use the Euler equation to find the solution to the brachistochrone problem.

5.7. The simplified control problem

Consider the problem

(5.26)
$$\frac{\partial T}{\partial t} = \frac{\partial^2 T}{\partial x^2} + f(x,t), \qquad 0 < x < \infty, \quad t > 0,$$

(5.27)
$$T(0,t) = h(t), \qquad t > 0 ,$$

(5.28)
$$T(x,0) = T_0(x), \qquad x > 0$$

with the additional condition that

(5.29)
$$\begin{array}{c} T(x,t) \text{ is bounded (as } x \to \infty) \\ \text{for any bounded time interval.} \end{array}$$

This problem can be solved with the aid of *Green's function*

$$G((x,t); y) = \frac{\exp(-\frac{(x-y)^2}{4t})}{\sqrt{4\pi t}} - \frac{\exp(-\frac{(x+y)^2}{4t})}{\sqrt{4\pi t}}$$

in the form

$$T(x,t) = \int\limits_0^t \int\limits_0^\infty G(x, t - \tau; t) f(y, \tau) dy d\tau$$

(5.30)

$$- \int\limits_0^t G_y(x, t - \tau; 0) h(\tau) d\tau + \int\limits_0^\infty G(x, t; y) T_0(y) dy.$$

PROBLEM 5.7.1. Prove that if f is a continuously differentiable, bounded function, and h, T_0 are continuous bounded functions, then any solution of (5.26)–(5.29) must have the form (5.30).

Hint. Use the proof of the representation formula stated below with K replaced by G.

Conversely, we can prove that the right-hand-side of (5.30) is a solution to (5.26)–(5.29).

PROBLEM 5.7.2. Prove this in the special case where $f \equiv 0$. (The case $f \not\equiv 0$ is more difficult.)

THEOREM 5.7.1. (Representation formula for the inhomogenous heat equation). *Let Ω denote the cylindrical region $x \in \omega$, $t \in (0, T)$, where ω is an open bounded region in n-dimensional Euclidean space $\mathbb{R}^n$ with C^2 (twice continuously differentiable) boundary $\partial\omega$. Let $u = u(x, t)$, u_t, $u_{x_i x_j}$ exist and be continuous in the closure $\overline{\Omega}$ of Ω and satisfy the equation*

$$u_t - \Delta u = f(x, t),$$

where $f(x, t)$ is a continuous function in $\overline{\Omega}$. Then, if $\xi \in \omega$, we have

$$u(\xi, T) = \int\limits_\Omega K(x, \xi, T - t) f(x, t) dx dt + \int\limits_{x \in \omega} K(x, \xi, T) u(x, 0) dx$$

(5.31)

$$+ \int\limits_0^T dt \int\limits_{x \in \partial\omega} \left(K(x, \xi, T-t) \frac{du(x, t)}{dn_x} - u(x, t) \frac{dK(x, \xi, T - t)}{dn_x} \right) dS_x.$$

Here $K(x, \xi, t)$ is the fundamental solution of the heat equation:

$$K(x, \xi, t) = \frac{\exp(-\frac{|x - \xi|^2}{4t})}{(4\pi t)^{n/2}}$$

and d/dn_x is the derivative in the direction of the exterior normal n_x of $\partial\omega$ at x.

Proof. For arbitrary $v(x,t) \in C^2(\overline{\Omega})$ we have, by Green's identity and integration by parts,

$$
\int_{\Omega} vf\,dx\,dt = \int_{\Omega} v(u_t - \Delta u)\,dx\,dt
$$

$$
= -\int_{\Omega} u(v_t + \Delta v)\,dx\,dt + \int_{\substack{x\in\omega\\t=T}} uv\,dx - \int_{\substack{x\in\omega\\t=0}} uv\,dx
$$

$$
-\int_0^T dt \int_{x\in\partial\omega} \left(v\frac{du}{dn} - u\frac{dv}{dn} \right) dS_x.
$$

Then, for a fixed $\xi \in \omega$ and $\varepsilon > 0$ choose $v(x,t) = K(x,\xi,T+\varepsilon-t)$ so that $v_t + \Delta v = 0$.

For $\varepsilon \to 0$,

$$
\int_{\substack{x\in\omega\\t=T}} uv\,dx = \int_{x\in\omega} K(x,\xi,\varepsilon)u(x,T)\,dx \to u(\xi,T)
$$

since

$$
w(\xi,\varepsilon) = \int_{x\in\omega} K(x,\xi,\varepsilon)u(x,T)\,dx
$$

is a solution to $w_\varepsilon - \Delta_\xi w = 0$ with initial values $w(\xi,0) = u(x,T)$. Then, since $K(x,\xi,T+\varepsilon-t)$ is uniformly continuous in ε,x,t for $\varepsilon \geq 0$, $x \in \partial\omega$, $0 \leq t \leq T$ and for $x \in \omega$, $t = 0$, and since $d/dn_x K(x,\xi,T+\varepsilon-t)$ is uniformly continuous in ε,x,t for $\varepsilon \geq 0$, $x \in \partial\omega$, $0 \leq t \leq T$, the assertion (5.31) follows as we let $\varepsilon \to 0$.

Now consider the system (5.19)–(5.23). From (5.20) we get

$$
(5.32) \qquad u(x,t) = u_0 \exp\left(-\int_0^x \left(\frac{T}{1+T}\right)(y,t)\,dy \right).
$$

We can therefore write

$$
\frac{Tu}{1+T} = \Phi(T),
$$

where Φ is a "functional" of T; it depends nonlocally on T. If we substitute this into (5.19), and then use the representation (5.30), we find that T satisfies

$$
(5.33) \qquad T(x,t) = \int_0^t \int_0^\infty G(x,t-\tau;y)\Phi(T)(y,\tau)\,dy\,d\tau + k(x,t),
$$

where

(5.34) $$k(x,t) = -\int_0^t G_y(x, t - \tau; 0)S(\tau)d\tau.$$

Note that $k(x,t)$ is a known function (for a given control function S).

We have reduced the problem (5.19)–(5.23) to a single "integral equation" (5.33). Although the expression $\Phi(T)$ is nonlocal in T, the method of successive iteration (which was used for solving ordinary differential equations) can actually be used here too to prove that (5.33) has a unique solution. Indeed, defining a sequence T_n successively by

$$T_{n+1}(x,t) = \int_0^t \int_0^\infty G(x, t - \tau; y)\Phi(T_n)(y, \tau)dyd\tau + k(x,t),$$

we can show that the sequence is convergent to a solution. The same method can also be used to prove uniqueness. Prove uniqueness!

To solve (5.19)–(5.23) numerically we can use either explicit schemes or implicit schemes.

1. One *implicit scheme* is

$$\frac{T_m^{n+1} - T_m^n}{\Delta t} = \frac{T_{m-1}^{n+1} - 2T_m^{n+1} + T_{m+1}^{n+1}}{\Delta x^2} + \frac{T_m^n}{1 + T_m^n}u_m^n,$$

$$\frac{u_{m+1}^n - u_m^n}{\Delta x} = -\frac{T_m^n}{1 + T_m^n}\,u_m^n.$$

2. One *explicit* scheme is

$$\frac{T_m^{n+1} - T_m^n}{\Delta t} = \frac{T_{m-1}^n - 2T_m^n + T_{m+1}^n}{\Delta x^2} + \frac{T_m^n}{1 + T_m^n}\,u_m^n,$$

$$\frac{u_{m+1}^n - u_m^n}{\Delta x} = -\frac{T_m^n}{1 + T_m^n}\,u_m^n.$$

PROBLEM 5.7.3. Take $L = 10$, $t_0 = 4$ and two choices of $S(t)$:

$$S(t) = 1 \quad \text{if } 0 \le t \le 4$$

and

$$S(t) = \begin{cases} \frac{1}{2} & \text{if } 0 \le t \le 2 \\ \frac{3}{2} & \text{if } 2 \le t \le 4. \end{cases}$$

For which of them is $J(S)$ smaller? Use each of the numerical methods described above and compare. Take the length of the converter to be either L or $2L$ and impose the boundary condition $T_x = 0$ at the right endpoint of the converter.

The ideas from control theory described in preceding sections can be applied in principle to the present problem. However, our model is still too complicated, and the conclusions that can be derived do not give sufficient insight. In the next section we further simplify the model and then use the ideas of optimal control theory in order to actually determine explicitly the optimal control.

5.8. Determining the optimal control

In this section we replace (5.19) by the heat equation

$$(5.35) \qquad \frac{\partial T}{\partial t} = \frac{\partial^2 T}{\partial x^2}, \qquad 0 < x < \infty, \quad 0 < t < T.$$

The underlying assumption here is that Tu is small compared to either $\partial t/\partial t$ or $\partial^2 T/\partial x^2$. We also assume, as in (5.24), that $T(x,t) \to 0$ if $x \to 0$. Then, by the representation formula (5.30),

$$(5.36) \qquad T(x,t) = \int_0^t \frac{x}{\sqrt{4\pi}(t-\tau)^{3/2}} \exp\left(-\frac{x^2}{4(t-\tau)}\right) S(\tau)d\tau.$$

Next we simplify the functional $J(S)$ by replacing

$$u_0 \int_0^{t_0} \exp\left(-\int_0^L \frac{T(y,t)}{1+T(y,t)}dy\right) dt \quad \text{by} \quad u_0 \int_0^{t_0} \left[1 - \int_0^\infty T(y,t)dy\right] dt.$$

This means that we have taken $L = \infty$, that

$$A \equiv \int_0^\infty \frac{T(y,t)}{1+T(y,t)}dy$$

is small so that $e^{-A} \approx 1 - A$, and that $T \ll 1$ so that $1 + T \approx 1$. Setting

$$(5.37) \qquad J_0(S) = u_0 \int_0^{t_0} \int_0^\infty T(y,t)dy\,dt,$$

the optimal control problem reduces to the following:

$$(5.38) \qquad \underset{S \in \mathcal{A}}{\text{maximize}}\, J_0(S).$$

We assume that
$$(5.39) \qquad Nt_0 < M.$$

This means that there is not enough "energy" M to keep the temperature at $x = 0$ at the maximal possible level N for all time t, $0 < t \leq t_0$. Thus we must use the available energy "wisely."

THEOREM 5.8.1. *There is a unique optimal solution $S_0(t)$ of* (5.38) *and it is given by*

$$(5.40) \qquad\qquad S_0(t) = \begin{cases} N & \text{if } 0 \leq t \leq \tilde{t} \\ 0 & \text{if } \tilde{t} < t < t_0, \end{cases}$$

where $\tilde{t}$ is determined by $N\tilde{t} = M$.

Proof. If we substitute T from (5.36) into (5.37), we get

$$
\begin{aligned}
J_0(S) &= u_0 \int_0^{t_0} dt \int_0^t S(\tau) d\tau \int_0^{\infty} \frac{x}{\sqrt{4\pi}} \frac{\exp\left(-\frac{x^2}{4(t-\tau)}\right)}{(t-\tau)^{3/2}} dx \\
&= \frac{2u_0}{\sqrt{4\pi}} \int_0^{t_0} dt \int_0^t S(\tau) \left[\frac{\exp\left(-\frac{x^2}{4(t-\tau)}\right)}{\sqrt{t-\tau}} \right]_{x=0}^{x=\infty} d\tau \\
&= \frac{2u_0}{\sqrt{4\pi}} \int_0^{t_0} S(\tau) d\tau \int_\tau^{t_0} \frac{dt}{\sqrt{t-\tau}} = \frac{4u_0}{\sqrt{4\pi}} \int_0^{t_0} \sqrt{t_0-\tau}\, S(\tau) d\tau.
\end{aligned}
$$

Since $\sqrt{t_0 - \tau}$ is strictly monotone decreasing in τ, the last integral is maximized if and only if $S(\tau)$ is chosen as in (5.40). (Verify it!)

We have completely characterized the optimal control for the functional $J_0(S)$. For more complicated functionals, we can still get some information on optimal controls by using ideas from the calculus of variations. Consider the functional

$$(5.41) \qquad\qquad J_1(S) = u_0 \int_0^{t_0} \exp\left(-\int_0^L T(y,t) dy \right),$$

where T is defined by (5.36).

THEOREM 5.8.2. *If $S_0(t)$ is a maximizer of $J_1(S)$ in the class $\mathcal{A}$ then $S_0(t)$ must take values 0 or N at some time t, $0 \leq t \leq t_0$.*

Proof. We suppose that

$$0 < S_0(t) < N \quad \text{for all } 0 \leq t \leq t_0$$

and derive a contradiction. Let $\zeta(t)$ be any continuous function in $0 \leq t \leq t_0$ such that

$$(5.42) \qquad\qquad \int_0^{t_0} \zeta(t) dt = 0.$$

Our assumption on $S_0(t)$ implies that $S_0 + \varepsilon\zeta$ is in $\mathcal{A}$ for all ε small enough. Denote by T_0 and T_1 the solutions (defined as in (5.36)) corresponding to S_0 and ζ, respectively. Then

$$(5.43) \qquad T_1(x,t) = \frac{1}{\sqrt{4\pi}} \int_0^t \frac{x}{(t-\tau)^{3/2}} \exp\left(-\frac{x^2}{4(t-\tau)}\right) \zeta(\tau)d\tau.$$

We now write

$$J(S_0 + \varepsilon\zeta) = \int_0^{t_0} u_0 \exp\left(\int_0^L [T_0(y,t) + \varepsilon T_1(y,t)] \, dy\right) dt \equiv I(\varepsilon).$$

Then $I(\varepsilon)$ takes minimum at $\varepsilon = 0$, and therefore $I'(0) = 0$. This implies that

$$\int_0^{t_0} \exp\left(-\int_0^L T_0(y,t)dy\right) dt \int_0^L T_1(y,t)dy = 0.$$

Substituting T_1 from (5.43) and proceeding similarly to the computation in the proof of Theorem 5.8.1, we obtain

$$\int_0^{t_0} \exp\left(-\int_0^L T_0(y,t)dy\right) dt \int_0^t \frac{1 - \exp\left(-\frac{L^2}{4(t-\tau)}\right)}{\sqrt{t-\tau}} \zeta(\tau)d\tau = 0$$

or

$$(5.44) \qquad \int_0^{t_0} f(\tau)\zeta(\tau) = 0,$$

where

$$(5.45) \qquad f(\tau) = \int_\tau^{t_0} \exp\left(-\int_0^L T_0(y,t)dy\right) \frac{1 - \exp\left(-\frac{L^2}{4(t-\tau)}\right)}{\sqrt{t-\tau}} dt.$$

It can be shown that

(5.46) if (5.44) holds for any continuous function ζ satisfying (5.42), where $f(\tau)$ is a given continuous function, then $f(\tau) = $ const.

Consequently, the function $f(\tau)$ given by (5.45) satisfies

$$f(\tau) = C, \qquad C \text{ constant.}$$

Since $f(\tau) \to 0$ if $t \to t_0$, C is equal to 0. On the other hand, if $0 \le \tau < t_0$, the integral $f(\tau)$ is positive (since the integrand is positive), which is a contradiction.

PROBLEM 5.8.1. Prove the assertion (5.46).

Hint. If $f(\tau_1) > f(\tau_2)$ choose ζ as in (5.42), which vanishes if $|\tau - \tau_1| > \varepsilon$ and $|\tau - \tau_2| > \varepsilon$ for ε small.

PROBLEM 5.8.2. Improve Theorem 5.8.2 by showing that, for any $0 < t_1 < t_0$, $S_0(t)$ must take values 0 or N in the interval $t_1 \leq t \leq t_0$. Try to show also that $S_0(t_0 - 0) = 0$.

PROBLEM 5.8.3. Extend Theorem 5.8.2 to the function $J(S)$, with T as in (5.36).

PROBLEM 5.8.4. Is the function $S_0(t)$ defined by (5.40) a solution to the optimization problem (5.41)? Explore the answer by numerical experiments.

Do the same for problem (5.19)–(5.25).

In the above study we have taken the catalytic converter to extend to $0 < x < \infty$. It is also natural to take the length of the catalytic converter to be finite, and in fact of length L. We then impose a boundary condition

$$(5.47) \qquad \frac{\partial T_s}{\partial x} + \beta T_s = 0 \quad \text{at } x = L \qquad (\beta > 0).$$

PROBLEM 5.8.5. Extend the explorations of Problem 5.8.4 to the case when T satisfies (5.47).

5.9. Summary

The physical and chemical principles of how a catalytic converter works are described. A mathematical model similar in many ways to the system of reaction diffusion equations of Chapter 4 was introduced in §5.2. The optimal control problem of minimizing the cost function $J(S)$ by choosing the optimal control function $S(t)$ was presented in §5.4. The optimal control is the one that minimizes the output of hazardous gases through the catalytic converter. A simplified system (5.19)–(5.23) was then derived for which the *direct* problem can be solved by finite differences (§5.7). Thus by trial and error certain choices of S can be seen to be better than others. Using ideas from the calculus of variations introduced in §5.6 and after some simplifying assumptions, properties of the optimal control problem are obtained in §5.8. In the most simplified model the optimal control $S_0(t)$ is uniquely determined by (5.40).

References

The model (5.5)–(5.10) was developed in

[1] S. H. Oh, J. C. Cavendish, and L. L. Hegedus, *Mathematical modeling of catalytic converter lightoff: Single-pellet studies*, AIChE J., 26 (1980),

pp. 935–943.

[2] S. H. Oh and J. C. Cavendish, *Mathematical modeling of catalytic converter lightoff. Part* II: *Model verification by engine-dynamometer experiments; Part* III: *Prediction of vehicle exhaust emissions and parametric analysis*, AIChE J., 31 (1985), pp. 935–942; pp. 943–949.

See also

[3] A. Friedman, *Mathematics in Industrial Problems, Part* 4, Chap. 7, IMA Vol. Math. Appl., Vol. 38, Springer–Verlag, New York, 1991.

Chapter 6

The Photocopy Machine

6.1. Background

When you take a picture with a camera, light penetrates through the shutter and creates an impression on the film. Later, the film is developed by putting it in a solution, whereby a chemical process takes place that transforms the "impression" on the film into a visible film negative, from which one then makes prints.

When you take a photocopy of a document, an entirely different process takes place. Instead of "light plus chemistry" as in pictures taken by camera, we now have "light plus electric charges." The photocopying process, which is referred to as electrophotography, is a much faster process than traditional photography (even when taking photos with instant cameras). In this chapter we explain how electrophotography works and study, in particular, the mathematical model involved in getting the electric image of the document (which is analogous to the "impression" that light makes on the film inside the camera). In the next chapter we study the mathematical model of the development of the visible image from the electric image (which is analogous to developing the film negative).

As we will see, the electric image is typically a blurred impression of the document, and the visible image is always an imprecise representation of the electric image. Mathematical analysis will show that this imprecision is inherent in the physical processes. Thus a photocopy can never be an absolutely precise copy of the document.

6.2. The photocopy machine

Inside the photocopier there is an aluminum drum of radius approximately two inches; it is called the photoconductor drum. Most of the photocopying process takes place around the photoconductor surface. This surface is also called a photoreceptor. The surface of the photoconductor is covered by two layers:

The inner layer, about 2.5μ (μ = micron) thick, is called the *generating layer*, and the outer layer of about 15μ is called the *transport layer*. Both layers are made of polymeric material and are solid.

The photocopying process is described in Fig. 6.1. It can be divided into seven steps.

Step 1. Charge corona. The device shown in Fig. 6.1 consists of hot wires that make the air ionize. The positive charges drop onto the surface of the photoconductor, just below the generating layer. The negative charges remain above the surface. The corona is a general phenomenon that occurs in the atmosphere and is seen around and close to a luminous body like the sun or moon; it is caused by the diffraction produced by droplets or particles of dust. It is also seen as a faint glow adjacent to the surfaces of electrical conductors such as electric poles with high voltage. When Step 1 is completed the machine is ready to begin photocopying.

Step 2. Charge. You have pushed the "start" bottom and green light begins to scan the document. The scanning light is reflected from the document and goes through additional reflection by one or several mirrors; it then passes through a lens and onto the surface of the photoconductor drum. The scanning light and the set of mirrors and lens are traveling together, and their motion is synchronized with the motion of the photoconductor drum; see Fig. 6.2.

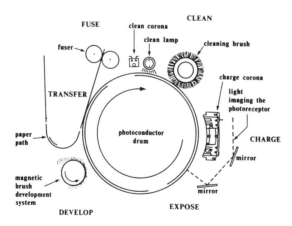

FIG. 6.1. *The photocopying process.*[1]

Step 3. Expose. The light incident to the photoconductor drum is converted into an electrical signal. This process will be modeled later in this

[1]A. Friedman, *Mathematics in Industrial Problems*, IMA Vol. Math. Appl., Vol. 24, Springer-Verlag, New York, 1989, Fig. 17.1, p. 157. ©1989, Springer-Verlag. Reprinted with permission.

chapter.

Step 4. Develop. The electric image is transformed into a visible image. This process will be modeled in Chapter 7.

Step 5. Transfer. The visible image is a spread of dry ink called toner. The darker spots correspond to the darker spots in the original document. Paper is fed with the aid of rollers, and it rubs against the surface of the photoconductor drum, becoming dark at the spots that correspond to the dark spots on the photoconductor surface.

Step 6. Fuse. To attach the visible image permanently onto the paper, the paper is transferred through hot rollers that fuse the image.

Step 7. Clean. A photocopy has been made but there is still "dirt" left on the photoconductor drum: It comes from the initial charge, the electric image, and the leftover toner. All this is cleaned, and the machine is then ready to make another copy.

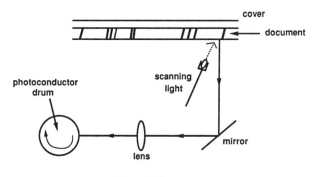

FIG. 6.2.

When you wish to start a cold photocopier, you must wait until the machine "warms up." This means that the fuse station has been made warm enough and the charge corona is in place.

6.3. The electric image

Figure 6.3 shows the transport layer, the generating layer, and the ground layer (i.e., the photoconductor drum). These layers are actually circular; taking them flat is a reasonable approximation.

Photons from the exposed light travel through the transport layer and collide with atoms in the generating layer. They impart energy to some of the electrons, which releases them from the pulling force of the atoms' nuclei. An atom that has lost such an electron is called a "hole."

The surface of the photoconductor drum is initially charged electrostatically (charge corona) with negative charge on top of the transport layer and positive charge in the ground layer. (see Step 1, explained in §6.2).

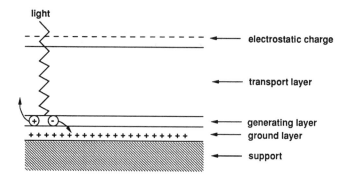

FIG. 6.3. *The photoconductor drum.*[2]

The electrons released in the generating layer are attracted to the positively charged ground layer. The holes, on the other hand, are attracted by the negative electrostatic field, and they move upward in the transport layer. This is described in Fig. 6.4.

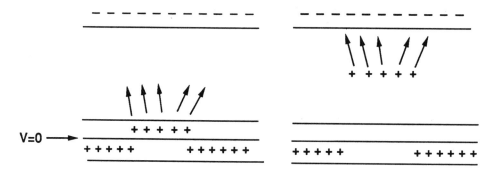

FIG. 6.4. (a) *Initial configuration.* (b) *Transient configuration.*[3]

6.4. Modeling the electric image

For simplicity we consider only the two-dimensional problem. Furthermore, we concentrate on a very simple document as depicted in Fig. 6.5. Since photons will not be reflected from the black part of the document, the density function of the holes in the generating layer at the initial time $t = 0$ has the profile described in Fig. 6.6.

---
[2]A. Friedman, *Mathematics in Industrial Problems*, IMA Vol. Math. Appl., Vol. 24, Springer-Verlag, New York, 1989, Fig. 17.3, p. 158. ©1989, Springer-Verlag. Reprinted with permission.

[3]A. Friedman, *Mathematics in Industrial Problems*, IMA Vol. Math. Appl., Vol. 24, Springer-Verlag, New York, 1989, Fig. 17.4, p. 159. ©1989, Springer-Verlag. Reprinted with permission.

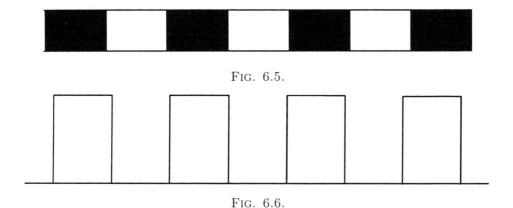

FIG. 6.5.

FIG. 6.6.

This is a periodic array, and we will concentrate on one period, say $0 \leq x \leq a$. Set

$$\Omega_0 = \{0 < x < a, h < y < \infty\} \quad \text{(air)},$$

$$\Omega_1 = \{0 < x < a, 0 < y < h\} \quad \text{(transport layer)},$$

$$\Omega_2 = \{0 < x < a, -\delta < y < 0\} \quad \text{(generating layer)},$$

and

$$\Omega = \{0 < x < a, \ -\delta < y < \infty, \ y \neq h\}$$

Introduce the dielectric constants

$$\kappa = \begin{cases} 1 & \text{in} \quad \Omega_0 \\ \kappa_1 & \text{in} \quad \Omega_1 \\ \kappa_2 & \text{in} \quad \Omega_2, \end{cases}$$

and set

$$\rho = \text{density of holes},$$

$$\mu_0 = \text{mobility of holes},$$

$$\vec{J} = \text{current density of holes},$$

$$\vec{E} = \text{electric field} = -\nabla V,$$

where V is the electric field potential (also called the voltage).

Then $\rho \equiv 0$ in Ω_0, and

$$(6.1) \qquad \frac{\partial \rho}{\partial t} = -\nabla \cdot \vec{J} \quad \text{in} \quad \Omega \quad \text{(continuity equation)},$$

where the flux $\vec{J}$ is given by

(6.2) $$\vec{J} = \mu_0 \rho \, \vec{E}$$

and

(6.3) $$\nabla \cdot (\kappa \, \vec{E}) = \frac{4\pi\rho}{\varepsilon_0} \quad \text{in } \Omega,$$

where ε_0 is the permitivity of the vacuum; $\varepsilon_0 = 8.85 \times 10^{-12}$ micro coulomb (μC). The last two equations follow from the Maxwell equations, which hold for general electromagnetic fields.

Using (6.2) in (6.1) and recalling that $\vec{E} = -\nabla V$, we obtain

(6.4) $$\frac{\partial \rho}{\partial t} = \mu_0 \nabla \cdot (\rho \nabla V) = \mu_0 (\nabla \rho) \cdot \nabla V + \mu_0 \rho \Delta V \quad \text{in } \Omega.$$

Equation (6.3) can also be written in the form

$$\nabla \cdot (\kappa \nabla V) = -\frac{4\pi\rho}{\varepsilon_0} \quad \text{in } \Omega.$$

Since κ_1 is actually very close to κ_2, we henceforth take

$$\kappa_1 = \kappa_2.$$

Then V satisfies the (so-called) Poisson equation

(6.5) $$\Delta V = -\frac{4\pi\rho}{\varepsilon_0 \kappa} \quad \text{in } \Omega.$$

Recall that $\rho \equiv 0$ in Ω_0.

Observe that if V is already known, then (6.4) is a first-order partial differential equation for ρ, similar to the advection equation in Chapter 2.

The holes are deposited at the top of the transport layer, i.e., at $\{y = h\}$. Let us denote their density on $\{y = h\}$ (including those that are initially there) by σ. Then, again from the Maxwell equations,

(6.6) $$[\kappa \partial_y V(x, y, t)]_{y=h-0}^{y=h+0} = -\frac{4\pi\sigma}{\varepsilon_0},$$

the square brackets denoting the jump across $y = h$. The function V is continuous across $y = h$, but its y-derivative is not.

The surface density σ increases by the current density of holes deposited at $y = h - 0$; according to (6.2) this is given by

(6.7) $$\frac{\partial \sigma}{\partial t} = -\mu_0 \rho \partial_y V \quad \text{at } y = h - 0.$$

Finally, we need to assume boundary conditions:

$$V = 0 \quad \text{on } y = -\delta,$$

(6.8)
$$V_y \to 0 \quad \text{if } y \to +\infty,$$

$$\rho(x, -\delta + 0, t) = 0,$$

$$V, \rho \quad \text{are periodic in } x, \text{ that is,}$$

$$V(0, y, t) = V(a, y, t),$$

(6.9)
$$V_x(0, y, t) = V_x(a, y, t),$$

$$\rho(0, y, t) = \rho(a, y, t),$$

and initial conditions:

(6.10)
$$\sigma(x, 0) = \sigma_0,$$

(6.11)
$$\rho(x, y, 0) = f(x, y),$$

where f vanishes for $y > 0$ and is a step function of x for $-\delta < y < 0$, say

$$f(x, y) = \begin{cases} A & \text{if } \alpha < x < \beta, \quad -\delta < y < 0 \\ B & \text{elsewhere in} \quad -\delta < y < 0, \end{cases}$$

where $0 < \alpha < \beta < a$. The initial charge σ_0 is due to the electrostatic charge; it is a negative number.

To solve (6.4)–(6.11), we extend all the initial data and the boundary conditions on $y = -\delta$, $y = h$, and $y \to \infty$ to all x by periodicity; for example, we define

$$f(x + a, y) = f(x, y) \quad \text{for all } x.$$

Then we seek to solve (6.4)–(6.8), (6.10), (6.11) for all x; we expect that the solution will be periodic so that, in fact, (6.9) will also be satisfied.

6.5. Solving Poisson's equations numerically

Suppose we want to solve

(6.12)
$$\Delta U = E(x) \quad \text{for } x = (x_1, x_2) \in \Omega,$$

(6.13)
$$U = 0 \quad \text{on the boundary } \partial\Omega \text{ of } \Omega.$$

Unfortunately, in higher dimensions, unlike in the one-dimensional case (where one deals with ordinary differential equations), boundary value problems cannot, in general, be solved explicitly by using *finite* linear combinations of explicit solutions. Yet two methods that we have already discussed come to mind, and for a simple region like a rectangle, they are quite useful. The first method is the use of Fourier series and separation of variables; the second is the method of finite differences.

Let us briefly describe the first method. To simplify matters, we assume the rectangle in question is given by

$$0 < x_1 < a, \qquad 0 < x_2 < b.$$

Then if $E(x_1, x_2)$ is continuous or piecewise continuous, it has a double sine Fourier series

$$E(x_1, x_2) \sim \sum_{p,q=1}^{\infty} E_{p,q} \sin \frac{p\pi x_1}{a} \sin \frac{q\pi x_2}{b},$$

which may or may not converge at every point (x_1, x_2) but does converge in an *average* sense. The sign "$\sim$" may be read as "has the Fourier series." One then looks for a solution U of the form

$$U(x_1, x_2) \sim \sum_{p,q=1}^{\infty} U_{p,q} \sin \frac{p\pi x_1}{a} \sin \frac{q\pi x_2}{b}.$$

Substituting formally gives us the relation

$$-\pi^2 \left(\frac{p^2}{a^2} + \frac{q^2}{b^2} \right) U_{p,q} = E_{p,q},$$

or

$$U(x_1, x_2) \sim -\frac{1}{\pi^2} \sum_{p,q=1}^{\infty} \frac{E_{p,q}}{\frac{p^2}{a^2} + \frac{q^2}{b^2}} \sin \frac{p\pi x_1}{a} \sin \frac{q\pi x_2}{b}.$$

Since each term in the series satisfies the zero boundary conditions, it is to be hoped (if one is lucky) that the infinite sum does also. If one terminates the series after N terms, the partial sum will satisfy the null boundary conditions and will be an approximate solution to the nonhomogeneous partial differential equation (6.12).

How does one compute the coefficients $E_{p,q}$? They are simply the Fourier sine coefficients and are given by

$$E_{p,q} = \frac{4}{ab} \int_0^a \int_0^b E(x_1, x_2) \sin \frac{p\pi x_1}{a} \sin \frac{q\pi x_2}{b} dx_1 dx_2.$$

PROBLEM 6.5.1. In the square

$$0 < x_1 < 10, \qquad 0 < x_2 < 10,$$

let $E(x_1, x_2)$ be the characteristic function of the set

$$L = \{1 \le x_1 \le 2 \,,\, 1 \le x_2 \le 9\} \cup \{1 \le x_1 \le 9 \,,\, 1 \le x_2 \le 2\}$$

describing a "fat" letter "L." Find $U(x_1, x_2)$ using double Fourier series. Use your judgement in how many terms to take in the double series.

The method of finite differences used earlier to investigate time-dependent problems can also be used successfully to treat time-independent problems. Thus let us consider (changing notation slightly)

(6.14) $$u_{xx} + u_{yy} = f(x, y) \quad \text{in } \Omega,$$

(6.15) $$u = g \qquad \text{in } \partial\Omega,$$

where $\partial\Omega$ is the boundary of Ω. Again suppose Ω is a rectangle (although this is not really necessary). We introduce meshpoints such that the distance to the nearest neighbor is $\Delta x = h$ and $\Delta y = h$. Then the Poisson equation (6.14) can be replaced by its finite difference analogue

(6.16) $$\frac{u_{m+1,n} - 2u_{m,n} + u_{m-1,n}}{h^2} + \frac{u_{m,n+1} - 2u_{m,n} + u_{m,n-1}}{h^2} = f_{m,n}$$

at *interior* mesh points, while

(6.17) $$u_{m,n} = g_{m,n}$$

at *boundary* mesh points (see Fig. 6.7).

Note that (6.16) can be replaced by

(6.18) $$u_{m,n} = \frac{1}{4}(u_{m+1,n} + u_{m,n+1} + u_{m-1,n} + u_{m,n-1}) - \frac{h^2}{4} f_{m,n},$$

which simply expresses the value $u_{m,n}$ as the average of its four nearest neighbors (see Fig. 6.8) minus $h^2/4 \, f_{m,n}$. Equation (6.18) together with (6.17) give us a system of N linear algebraic equations in N unknowns. N is the number of dots in Fig. 6.7, which in the particular case drawn is 56. The unknowns (the $u_{m,n}$) will appear on the left-hand sides of the equations, while the $f_{m,n}$ and the $g_{m,n}$ will appear on the right-hand sides.

The first question that arises is, is this system of N linear equations in N unknowns always solvable? This is equivalent to the question, does the corresponding homogeneous system only have the "trivial" or zero solution? But what is the corresponding homogeneous system? It is the system we get by

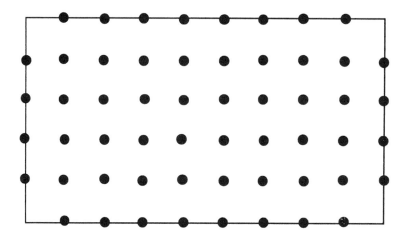

FIG. 6.7.

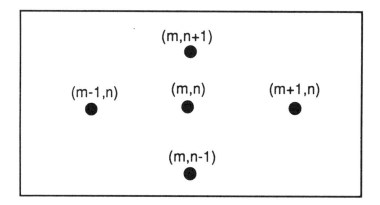

FIG. 6.8.

letting all the $f_{m,n}$ (in the interior) be zero, and all the $g_{m,n}$ (at the boundary) be zero.

In other words, if in the discretized problem, the equation is homogeneous, and the boundary data is zero, must the solution be zero? That the answer to this question is in the affirmative is due to the following "maximum principle."

THEOREM 6.5.1. (Maximum principle for the discrete Laplace equation). *Suppose at each interior point (m, n), $u_{m,n}$ is equal to the average of its four nearest neighbors. Then $u_{m,n}$ cannot have an interior maximum unless all the $u_{m,n}$ are equal to the same constant.*

Proof. Suppose the maximum is achieved at the interior point (m^*, n^*). Then the *average value* of its four nearest neighbors must be greater than or equal to *each one of them*. This can only happen if $u_{m,n}$ at (m^*, n^*) is equal to *each* of its four neighbors. Thus the maximum value is also attained by its four nearest neighbors. Applying the same argument to each of the neighbors, and repeating this process until all the points in the mesh are reached, we get the asserted conclusion.

Note that the same statement also holds for "minimum." Thus both the maximum and the minimum values of the $u_{m,n}$ must be taken on at the boundary mesh points. Therefore, if the $g_{m,n}$ are all zero, the maximum and minimum values of the $u_{m,n}$ must also be zero, and therefore the $u_{m,n}$ themselves must all be zero.

This shows that the discretized problem (6.18), (6.17) always has a unique solution.

The next question is one of convergence. Do the $u_{m,n}$ converge in some sense to "the" solution of the differential equation? Using the discrete maximum principle just proved, together with the known fact that the solution of the original (nondiscretized) problem has a unique solution (under appropriate smoothness conditions on f and g) it can be shown that the $u_{m,n}$ converge, as the mesh size $h \to 0$, to a function $u(x, y)$ that satisfies the differential equation (6.14) together with the boundary condition (6.15).

PROBLEM 6.5.2. Repeat Problem 6.5.1, but this time use finite differences.

REMARK 6.5.1. In §6.8 we describe an effective method to solve the problem (6.18), (6.17).

6.6. Transmission conditions

Consider the very simple ordinary differential equations $y'' = 0$ in the intervals $a < x < c$ and $c < x < b$. Then y and y' will have one-sided limits at $x = a, b, c$. To be a solution in the larger interval $a < x < b$, it is necessary and sufficient that $y(c - 0) = y(c + 0)$ and $y'(c - 0) = y'(c + 0)$. This is related to the fact that the values of y and y' at a point in general determine the solution of a second-order ordinary differential equation uniquely. Thus the fact just stated for the equation $y'' = 0$ holds more generally for second-order ordinary differential equations.

More generally one could require instead of the *homogeneous* jump conditions

$$[y(x)]_{x=c-0}^{x=c+0} = 0, \qquad [y'(x)]_{x=c-0}^{x=c+0} = 0$$

nonhomogeneous jump conditions

$$[y(x)]_{x=c-0}^{x=c+0} = h_0, \qquad [y'(x)]_{x=c-0}^{x=c+0} = h_1;$$

most commonly h_0 is taken to be zero. Such conditions are called *transmission conditions.* For example, the problem

$$v'' = f(x) \qquad a < x < b, \ x \neq c,$$

$$v(a) = A, \qquad v(b) = B,$$

$$v(c - 0) = v(c + 0),$$

$$v'(c - 0) - v'(c + 0) = C$$

contains two boundary conditions at a and b and a transmission condition at c. It is called a *transmission problem.* Physically $v(x)$ may be thought of as the voltage due to a continuous charge distribution $f(x)$ and a single charge at $x = c$ of magnitude C. A and B are the (known) voltages at points a and b.

Transmission conditions also make sense for *partial differential equations.* For example, suppose D is a two-dimensional domain bounded by a curve C on the "outside" and a line segment $\ell : a < x < b$, $y = h$ from the "inside." Then it can be shown, in analogy to the one-dimensional situation, that for a harmonic function $u(x, y)$ in D to be extendable to a harmonic function in $D \cup \ell$, it is necessary that the one-sided limits of u and u_y as $y \to h$ on ℓ from both sides of ℓ exist, and that

$$[u(x,y)]_{y=h-0}^{y=h+0} = 0, \qquad [u_y(x,y)]_{y=h-0}^{y=h+0} = 0$$

for $a \leq x \leq b$. (If instead of a straight line segment we had a curve, we would replace u_y by the normal derivative $\partial u / \partial n$ to the curve.)

Now instead of the *homogeneous* jump conditions, we could have inhomogeneous jump conditions, i.e., the zeros are replaced by functions of x, as in §6.4, (6.6). Then we would have *transmission* conditions again. Thus a typical transmission problem would be

$$\Delta u(x, y) = f(x, y) \quad \text{in} \ \ D,$$

$$u(x, y) = g(x, y) \quad \text{on} \ \ C,$$

$$[u(x,y)]_{y=h-0}^{y=h+0} = 0, \qquad [u_y(x,y)]_{y=h-0}^{y=h+0} = h(x) \quad \text{for} \ \ a \leq x \leq b.$$

Here, as in the one-dimensional case, u represents the potential (or voltage) due to a two-dimensional charge distribution of density $f(x, y)$ and a one-dimensional charge distribution on ℓ of density $h(x)$ if the potential on the exterior boundary C is known to be equal to $g(x, y)$.

6.7. Computing the electric image

The initial density of holes $f(x, y)$ reflects the impression of the document, generated by the photons. The profile of $\sigma(x, t) - \sigma_0$ is the electric image of f at time t. Figure 6.9 describes a typical shape of σ for large enough t. It shows the correlation of the electric image to f; the image is typically not a sharp reproduction of f. This can be proved mathematically and is also substantiated by calculations.

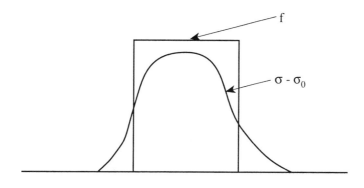

FIG. 6.9.

The following are typical values for constants introduced above:

$$a = 200\mu, \quad h = 15\mu, \quad \delta = 2.5\mu,$$

$$\kappa_1 = \kappa_2 = 3, \quad \mu_0 = 40,$$

$$\sigma_0 = -5 \times 10^{-10} \frac{\mu C}{\mu^2},$$

$$A = 177 \times 10^{-12} \frac{\mu C}{\mu^3}, \qquad B = 36 \times 10^{-12} \frac{\mu C}{\mu^3}.$$

The values α, β are fractions of a, e.g., $a/4, a/3, 2a/5$, etc. It is expected that within $\frac{1}{2}$ second the absolute value of $\sigma(t)$ will decrease to about $\frac{1}{200}$ of its original value.

PROBLEM 6.7.1. Solve (6.4)–(6.11) for $V = V(y)$, given $\sigma(x, t) = \sigma_0$ and $\rho(x, y, t) \equiv A$ for $-\delta < y < 0$ and zero otherwise. Find V explicitly in terms of A, σ_0, and compute it for the numerical values given above (with $B = 0$).

The system (6.4)–(6.11) can be attacked by finite differences. We first assume that the unknowns are *independent* of x (this will be the case if σ_0 and $\rho(x, y, 0)$ are independent of x. Note that (6.5) and (6.8) imply that V is constant for $y > h$, and hence (6.6) becomes

$$V_y(h - 0, t) = \frac{4\pi\sigma}{\varepsilon_0\kappa}.$$

We approximate the original system by the following difference scheme:

(i) $\quad\quad \dfrac{\rho_m^{n+1} - \rho_m^n}{\Delta t} = \mu_0 \dfrac{\rho_m^n - \rho_{m-1}^n}{\Delta y} \cdot \dfrac{V_m^n - V_{m-1}^n}{\Delta y} - \dfrac{4\pi\mu_0(\rho_m^n)^2}{\varepsilon_0\kappa}$

for $m \le m_h$, setting $\rho \equiv 0$ for $m > m_h$.

(ii) $\quad\quad \dfrac{V_{m-1}^n - 2V_m^n + V_{m+1}^n}{(\Delta y)^2} = -\dfrac{4\pi\rho_m^n}{\varepsilon_0\kappa}$

for $y \ne h$, i.e., $m \ne m_h$,

(iii) $\quad\quad \dfrac{V_{m_h}^n - V_{m_h-1}^n}{\Delta y} = \dfrac{4\pi\sigma^n}{\varepsilon_0\kappa} \quad\quad (y = h - 0),$

(iv) $\quad\quad \dfrac{\sigma^{n+1} - \sigma^n}{\Delta t} = -\mu_0\rho_{m_h}^n \dfrac{V_{m_h}^n - V_{m_h-1}^n}{\Delta y} \quad\quad (y = h - 0)$

$$= -\mu_0\rho_{m_h}^n \cdot \dfrac{4\pi\sigma^n}{\varepsilon_0\kappa} \quad\quad \text{(by (iii))}.$$

For every time step n, we perform the computation in the following order: First we solve (ii) and (iii) for the V_m's, using the boundary conditions; note that for $n = 0$ we use the initial conditions for ρ and σ. Then we solve (iv) for σ^{n+1} and finally solve (i) for the ρ_m^{n+1}'s. In solving (ii) and (iii) one may use an LU decomposition of the linear system with unknown vector $(V_1^n, V_2^n, \ldots, V_m^n)^T$.

PROBLEM 6.7.2. Carry through this procedure for (6.4)–(6.11) for the data listed before Problem 6.7.1. Let the final time be 0.01 second. Try $\Delta x = 2 \times 10^{-6}$, $\Delta t = 1 \times 10^{-4}$.

Actually there is some diffusion in the motion of the density of holes ρ. Thus (6.4) should be replaced by

(6.19) $\quad\quad\quad\quad \dfrac{\partial\rho}{\partial t} = \mu_0\nabla \cdot (\rho\nabla V) + \varepsilon\Delta\rho$

for some small $\varepsilon > 0$.

Equation (6.19) is similar to the heat equation. By the maximum principle,

$$\rho(x, y, t) \ge 0.$$

PROBLEM 6.7.3. Take $\alpha = a/4, \beta = 3a/4$. Devise a finite difference scheme for (6.4)–(6.11) analogous to the one described above, but this time allowing periodic x dependence. Replace the semi-infinite y-interval by a finite one $(-\delta, b)$; try $b = 35\mu$. Calculate V at $t = 0$. Then carry through the calculations one time step. Note that the analogue of (iii) must be changed.

PROBLEM 6.7.4. Carry through the computations of Problem 6.7.3 several time steps. Graph the profile of $\sigma(x, t_{\text{final}})$ for several values of σ_0; for example,

$$-5 \times 10^{-10} \frac{\mu C}{\mu^3}, \qquad -100 \times 10^{-10} \frac{\mu C}{\mu^3}.$$

Does the profile of σ become sharper as σ_0 increases?

PROBLEM 6.7.5. Solve Problems 6.7.3 and 6.7.4 when (6.4) is replaced by (6.19) taking $\varepsilon = 1, \frac{1}{10}$. How does the profile of $\sigma(x, t)$ compare with its profile in the case $\varepsilon = 0$?

Hint. Proceed by the method described in (i)–(iv) above. We need to include, however, finite differences also with respect to x as well as finite difference approximation to $\varepsilon \Delta \rho$.

Note 1. In solving the Poisson equation, try the methods of §6.8 or the implicit methods discussed in §4.6. Which is more effective?

Note 2. Problems 6.7.3–6.7.5 may be very time consuming. Perhaps a term paper on these problems might be appropriate.

6.8. A simple method for solving (6.18), (6.17)

A very simple method for obtaining the solution of the system of equations (6.18), (6.17) is an iterative method called the Jacobi method. Starting from values arbitrarily guessed, $\{u_{m,n}^{(0)}\}$, this method generates a sequence $\{u_{m,n}^{(1)}\}$, $\{u_{m,n}^{(2)}\}, \ldots$, that converges to the solution of our system. The method is the following. Letting u_0 be the initial guess, set

$$(6.20) \qquad \begin{cases} u_{m,n}^{(0)} = u_0(mh, nh) \text{ at the interior points} \\ \text{and } u_{m,n}^{(0)} = g_{m,n} \text{ at the boundary points.} \end{cases}$$

Knowing $\{u_{m,n}^{(k)}\}$, compute $\{u_{m,n}^{(k+1)}\}$ as follows:

$$(6.21) \qquad u_{m,n}^{(k+1)} = \tfrac{1}{4}\left(u_{m+1,n}^{(k)} + u_{m,n+1}^{(k)} + u_{m-1,n}^{(k)} + u_{m,n-1}^{(k)}\right) - \tfrac{1}{4}f_{m,n}h^2$$

at the interior points and $u_{m,n}^{(k+1)} = g_{m,n}$ at the boundary points. Of course, if $\{u_{m,n}^{(k)}\}$ is a good approximation to the solution of (6.18), (6.17), we do not need to compute $\{u_{m,n}^{(k+1)}\}$. To determine where it is reasonable to stop the computations, we must study the way in which $\{u_{m,n}^{(k)}\}$ approaches the exact solution $\{u_{m,n}\}$. Denote by $e_{m,n}^{(k)}$ the value at the meshpoint (mh, nh) of the error at the iteration number k, that is, $e_{m,n}^{(k)} = u_{m,n} - u_{m,n}^{(k)}$. Since $e_{m,n}^{(k)}$ is equal to zero at the boundary points, we can express $e_{m,n}^{(k)}$ as follows:

$$(6.22) \qquad e_{m,n}^{(k)} = \sum_{p=1}^{M_x} \sum_{q=1}^{M_y} C_{p,q}^{(k)} \sin\left(\frac{p\pi h}{a} m\right) \sin\left(\frac{q\pi h}{b} n\right),$$

where $M_x = (a/h) - 1$ and $M_y = (b/h) - 1$. Compare this expansion with the complete Fourier series in §6.5. Note that the higher frequencies p, q may be omitted because at the meshpoints the corresponding sine terms agree with those of the lower frequencies. By analogy with the case of a function $e(x, y)$ defined for every (x, y) in Ω, where the "norm" $\|e\|$ is defined by

$$\|e\|^2 = \iint\limits_{\Omega} |e(x,y)|^2 dx dy,$$

we set

$$\|e(k)\|^2 = \frac{ab}{4} \sum_{p=1}^{M_x} \sum_{q=1}^{M_y} |C_{p,q}^{(k)}|^2,$$

and we consider it as a measure of the size of the error at the iteration number k.

To find out how this error behaves as the iterations go on, we first note that combining (6.18) and (6.21), we can write that

$$(6.23) \qquad e_{m,n}^{(k+1)} = \tfrac{1}{4}(e_{m+1,n}^{(k)} + e_{m,n+1}^{(k)} + e_{m-1,n}^{(k)} + e_{m,n-1}^{(k)}).$$

Since

$$\sin\left(\frac{p\pi h}{a}(m+1)\right) + \sin\left(\frac{p\pi h}{a}(m-1)\right) = \left(2 - 4\sin^2\left(\frac{p\pi h}{2a}\right)\right)\sin\left(\frac{p\pi h}{a}m\right),$$

$$\sin\left(\frac{q\pi h}{b}(n+1)\right) + \sin\left(\frac{q\pi h}{b}(n-1)\right) = \left(2 - 4\sin^2\left(\frac{q\pi h}{2b}\right)\right)\sin\left(\frac{q\pi h}{b}n\right),$$

after inserting the expression (6.22) into (6.23), we obtain

$$e_{m,n}^{(k+1)} = \sum_{p=1}^{M_x} \sum_{q=1}^{M_y} s_{p,q} C_{p,q}^{(k)} \sin\left(\frac{p\pi h}{a}m\right)\sin\left(\frac{q\pi h}{b}n\right),$$

where

$$s_{p,q} = 1 - \sin^2\left(\frac{p\pi h}{2a}\right) - \sin^2\left(\frac{q\pi h}{2b}\right).$$

The size of $s_{p,q}$ tells us how the component of the error associated with the frequencies p, q is modified after one iteration. As a consequence,

$$\|e^{(k+1)}\|^2 = \frac{ab}{4} \sum_{p=1}^{M_x} \sum_{q=1}^{M_y} s_{p,q}^2 |C_{p,q}^k|^2$$

$$\leq \max_{\substack{1 \leq p \leq M_x \\ 1 \leq q \leq M_y}} |s_{p,q}|^2 \cdot \frac{ab}{4} \sum_{p=1}^{M_x} \sum_{q=1}^{M_y} |C_{p,q}^k|^2.$$

In other words,

$$\|e^{k+1}\| \leq \max_{\substack{1\leq p\leq M_x \\ 1\leq q\leq M_y}} |s_{p,q}| \cdot \|e^k\|$$

$$\leq \left[\max_{\substack{1\leq p\leq M_x \\ 1\leq q\leq M_y}} |s_{p,q}| \right]^{k+1} \|e^0\|.$$

Since

$$\max_{\substack{1\leq p\leq M_x \\ 1\leq q\leq M_y}} |s_{p,q}| = |s_{11}|$$

$$= |1 - \sin^2\left(\frac{\pi h}{2a}\right) - \sin^2\left(\frac{\pi h}{2b}\right)| < 1$$

the method does converge.

The above error analysis helps us to determine the number of iterations we need to do. If our initial error is order one, it is reasonable to stop the computations when it has been reduced, say, one million times. The number of iterations we need to do, k^*, can be computed as follows:

$$|s_{11}|^{k^*} \leq 10^{-6}$$

and so

$$k^* \geq -\frac{6\ln 10}{\ln |g_{11}|}.$$

For h very small,

$$|s_{11}| \approx 1 - \frac{\pi^2}{4}\left(\frac{1}{a^2} + \frac{1}{b^2}\right) h^2,$$

and so, the number of iterations we have to do is

$$k^* \geq \frac{24}{\pi^2} \cdot \ln 10 \cdot \frac{a^2 b^2}{a^2 + b^2} \, h^{-2}$$

$$= \frac{24}{\pi^2} \cdot \ln 10 \cdot \frac{M_x^2 \, M_y^2}{M_x^2 + M_y^2} \, .$$

PROBLEM 6.8.1. Show that the total number of operations of the Jacobi method (and hence the time it will take to run it on the computer) is proportional to h^4.

6.9. Summary

The physical principles of how a photocopy machine works were described. Seven steps of operation were briefly considered, the third of which, how the light signal gives rise to the electric image, was dealt with in this chapter. In §6.4 we introduced a coupled system for the time-dependent quantities ρ, V, σ. The density ρ satisfies an advection equation depending on V, and V satisfies a Poisson equation with right-hand side depending on ρ. This is further complicated by a transmission condition involving the surface charge density σ. In §6.5 we introduced two methods for solving Poisson's equation numerically. The second method results in a system (6.18) and (6.17) of linear algebraic equations, whose solution is addressed in §6.8. Schemes for solving the original system of partial differential equations and thus finding the electrical image $\sigma - \sigma_0$ were investigated.

References

An introduction to electrophotography is given in

[1] L. B. Schein, *Electrophotography and Development Physics*, Springer–Verlag, Heidelberg, 1988.

[2] D. M. Pai and B. E. Springett, *Physics of electrophotography*, Rev. Modern Phys., 65 (1993), pp. 163–211.

The model presented in this chapter was presented by John Spence in June 1989 and is described in

[3] A. Friedman, *Mathematics in Industrial Problems, Part* 2, Chap. 17, IMA Vol. Math. Appl., Vol. 24, Springer-Verlag, New York, 1989.

Existence theorems for the systems (6.4)–(6.11) were proved in

[4] A. Friedman and W. Liu, *A system of partial differential equations arising in electrophotography*, J. Differential Equations, 89 (1991), pp. 272–304.

Chapter 7

The Photocopy Machine (Continued)

In this chapter we model the development of visible image from the electric image.

7.1. The visible image

Figure 7.1 recalls how the electric image was formed.

It will be slightly more convenient to work with the function $u = -V$ rather than with the electric potential V. We also change notation of the surface hole's density, calling it $-\sigma$ instead of σ (and $-\sigma_0$ instead of σ_0, at time $t = 0$). Then, σ_0 is positive.

For simplicity we take the extreme levels of σ to be 0 and σ_0. We concentrate on one "cell" as in Fig. 7.1 with $-a < x < a$, say. We also assume that the electric image is sharp, that is, $\sigma = \sigma_0$ on an interval $-\varepsilon \leq x \leq \varepsilon$ that corresponds to the dark part of the document, and $\sigma = 0$ on the intervals $-a < x < -\varepsilon$ and $\varepsilon < x < a$ that correspond to the light part of the document. Thus σ is a step function and each step corresponds, roughly, to dark or light space in the document.

We now proceed to describe the development of the electric image into a visible image. This is accomplished by means of a positively charged chemical substance called a toner (something like a dry ink). The toner is transported

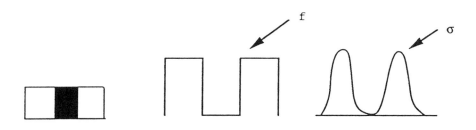

FIG. 7.1.

123

on the surface of carriers, which are iron balls of radius $\frac{1}{200}$ inch. The toner material flows into the surface of a roller, and under magnetic force, the toner carriers rearrange in chains or whiskers that brush against the photoconductor drum releasing some of their chemicals; see Fig. 7.2.

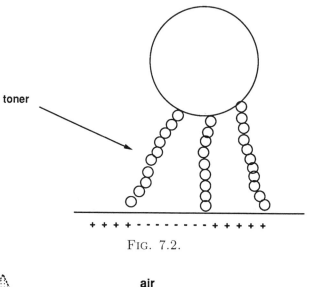

FIG. 7.2.

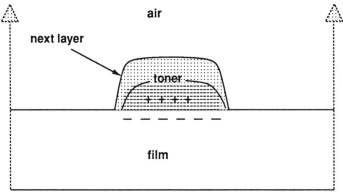

FIG. 7.3. [1]

As the charged surface of the photoconductor enters the development zone, i.e., the zone where the toner is deposited, visible dark images are created as the pattern on the charged surface becomes partially covered with toner. When the pattern has moved out of the development zone, it is "fully developed" with dark images. We wish now to concentrate on one cell and determine how it becomes fully developed. We adopt a "quasi-static" approach. In Fig. 7.3

---

[1]A. Friedman, *Mathematics in Industrial Problems*, IMA Vol. Math. Appl., Vol. 24, Springer-Verlag, New York, 1989, Fig. 17.7, p. 162. ©1989 Springer-Verlag.

the heavily shaded domain designates the toner image at present. In the next state another layer is added (lighter shaded domain).

Recall that after the visible image becomes fully developed, paper is fed onto the surface that supports the visible image, and the image is transferred to the paper. The paper then passes through a fuser that permanently attaches the image to the paper. The final step is the clean up of the photoconductor drum, preparing it to start the whole process again.

To describe the mathematical model of the development of the visible image, we must calculate the electric field, add toner proportional to the field strength in a layer adjacent to the previous layer, and keep iterating the process. The differential equation in the entire region (including the cell) is

(7.1)
$$\nabla \cdot (\kappa \vec{E}) = \frac{4\pi\rho}{\varepsilon_0}, \qquad \vec{E} = -\nabla V,$$

where ρ is the density of the toner (ρ constant), and the interface condition on the toner's boundary is

(7.2)
$$\vec{n} \cdot (\vec{E}_{\text{air}} - \kappa \vec{E}_{\text{toner}}) = 0,$$

where $\vec{n}$ is the normal to the interface, pointing in the direction of the air, and κ is the electric conductivity of the toner. The next layer is determined by the following rule:

$$\text{if } \vec{n} \cdot \vec{E}_{\text{air}} < 0, \text{ move out the interface,}$$

(7.3)

$$\text{if } \vec{n} \cdot \vec{E}_{\text{air}} > 0, \text{ move in the interface.}$$

If $\vec{n} \cdot \vec{E}_{\text{air}} = 0$ we do not move the interface since there is no force to attract or repel toner. It follows that the final position of the interface is determined by

(7.4)
$$\vec{n} \cdot \vec{E}_{\text{air}} = \vec{n} \cdot \kappa \vec{E}_{\text{toner}} = 0.$$

If we set $u = -V$, $k = 4\pi\rho/\varepsilon_0\kappa$, and impose a no-flux condition on the sides $x = \pm a$ of a cell, then the problem reduces to the following one.

Find a curve $\Gamma : y = f(x)$ $(-\beta \leq x \leq \beta)$ and a function $u(x,y)$, both symmetric in x, such that

(7.5)
$$f(x) > 0 \quad \text{if} \ -\beta < x < \beta, \qquad f(\pm\beta) = 0,$$

and, if

$$D = \{(x,y); -\beta < x < \beta \ , \ 0 < y < f(x)\},$$

$$\Omega = R \backslash \overline{D}, \qquad R = \{-a < x < a \ , -h < y < b\},$$

then

(7.6) $$\Delta u = k \quad \text{in} \quad D,$$

(7.7) $$\Delta u = 0 \quad \text{in} \quad \Omega,$$

(7.8) $$\frac{\partial u}{\partial y}(x, 0+) - \frac{\partial u}{\partial y}(x, 0-) = -\frac{4\pi\sigma_0}{\kappa\varepsilon_0} \equiv -\sigma_*, \qquad -\varepsilon < x < \varepsilon,$$

where σ_0 is a positive constant,

(7.9)
$$u \quad \text{is continuous in} \quad \overline{R},$$

$$u \quad \text{is continuously differentiable in} \quad R\backslash\{(x, 0); -\varepsilon \le x \le \varepsilon\},$$

(7.10)
$$u(x, b) = M, \quad -a < x < a,$$
$$u(x, -h) = 0, \quad -a < x < a,$$
$$u_x(\pm a, y) = 0, \quad -h < y < b,$$

and, finally,

(7.11) $$\frac{\partial u}{\partial n} = 0 \quad \text{on} \quad \Gamma = \{(x, f(x)); -\beta < x < \beta\} \cap R,$$

where n is the normal to Γ pointing upward.

Note that (7.8) is merely (6.6) with V replaced by u, σ renamed as $-\sigma$, and the simplifying assumption that σ is the constant σ_0 in the interval $-\varepsilon < x < \varepsilon$ and zero elswhere.

If Γ is given, the problem (7.6)–(7.10) is simply a transmission problem in R, with a transmission condition on the segment

$$\Sigma = \{(x, 0); -\varepsilon < x < \varepsilon\};$$

u is a solution of the equation

(7.12) $$\Delta u = k\chi_D \quad \text{in} \quad R\backslash\Sigma,$$

where χ_D is the characteristic function of D.

However Γ is *not* a priori known; it must be found, together with u, in such a way that the condition (7.11) is satisfied. Figure 7.4 summarizes the geometry of the problem.

This problem is an example of many other problems where a part of the boundary is not known and must be found together with the solution u to some differential equation with some boundary conditions. We call such a problem

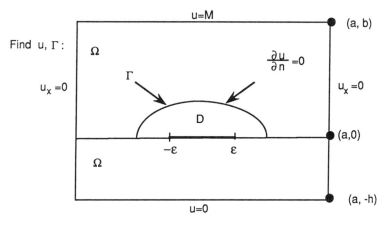

Find u, Γ :

FIG. 7.4.

a *free boundary problem*; the unknown part Γ of boundary is called the *free boundary*.

There exist two solutions to the toner problem with no free boundary. The first one is with $D = \{-a < x < a \, , \; 0 < y < b\}$ and the second one with $D = \emptyset$. We denote them by U_1 and U_0, respectively. These solutions correspond to visible images that are totally dark or totally light and therefore are not physical, except in certain extreme cases.

In the actual photocopier the various constants defined above have the following typical values:

$$a = 100\mu, \quad b = 600\mu, \quad h = 20\mu,$$

$$\sigma_0 = 4 \times 10^{-10}\frac{\mu C}{\mu^2}, \quad \kappa = 3, \quad \rho = 50 \times 10^{-12}\,\frac{\mu C}{\mu^3}$$

and then

$$\sigma_* = \frac{4\pi\sigma_0}{\kappa\varepsilon_0} = 182\frac{\mu C}{\mu^3}, \qquad k = 3.$$

Since also $M \sim 50V$, $M < \sigma_* h$ (i.e., $50 < 182 \times 20$).

In the following problem we assume that $\varepsilon = a$, i.e., the electric image is that of a totally dark cell.

PROBLEM 7.1.1. If $\varepsilon = a$, find explicitly the solutions U_1, U_0. (Assume they depend on y only.)

PROBLEM 7.1.2. Let $\varepsilon = a$. If $M > \sigma_* h$ then there does not exist a solution $u(y)$ of the toner problem with nonempty free boundary. If $M < \sigma_* h$ then there exists a solution to the toner problem with nonempty free boundry, provided $\sigma_* h - M < \frac{1}{2}\, b^2$; what happens when $\sigma_* h - M \geq \frac{1}{2}\, b^2$?

7.2. Impossibility of a precise image

If D in the previous section were a rectangle

$$(7.13) \qquad\qquad \{-\delta < x < \delta,\ 0 < y < \gamma\},$$

where $\delta = \varepsilon$, then we could say that the visible image gives a precise representation of the electrical image. According to the following theorem, this cannot happen.

THEOREM 7.2.1. *D cannot have the form (7.13) for any $0 < \delta < a$.*

Before proving this theorem we need some facts about harmonic functions. The first fact deals with the existence and uniqueness of the "Dirichlet problem."

THEOREM 7.2.2. *Let Ω be a disk or a rectangle with boundary $\partial\Omega$. Then given a continuous function $f(x,y)$ defined on $\partial\Omega$ there exists a unique continous function $u(x,y)$ in $\overline{\Omega} = \Omega \cup \partial\Omega$ which is in $C^2(\Omega)$ such that*

$$u_{xx} + u_{yy} = 0 \quad in\ \Omega$$

and $u = f$ on $\partial\Omega$.

The uniqueness part follows from the maximum principle. We omit the proof of existence.

The theorem holds for much more general Ω. For example Ω could be any bounded domain with boundary consisting of a finite number of curves smooth (C^1) except for a finite number of "corners."

Under what conditions can a solution to the "Dirichlet problem" be extended *past* the boundary $\partial\Omega$ into a larger domain? Since f can be an arbitrary *continuous* function, and hence can be chosen not to be smooth (say C^2) it is clear that *in general* the solution u cannot be extended *past* $\partial\Omega$ to be a (C^2) solution of Laplace's equation. However *in special circumstances* such an extension is possible, and that is a very useful fact. In what follows a bar will denote the closure of a set, e.g., $\overline{R_+}$ = closure of R_+.

THEOREM 7.2.3. (The Schwarz Reflection Principle). *Let R_+ be the rectangle*

$$\{a < x < b,\ 0 < y < c\}$$

and R_- its mirror image (see Fig. 7.5)

$$\{a < x < b,\ -c < y < 0\}.$$

If $u(x,y)$ satisfies $u_{xx} + u_{yy} = 0$ in R_+ and is continuous in $\overline{R_+}$, with $u = 0$ on $\ell \equiv \overline{R_+} \cap \overline{R_-}$, then u can be continued into $R \equiv R_+ \cup R_- \cup \ell$ as a harmonic function, by the formula

$$u(x,y) = -u(x,-y).$$

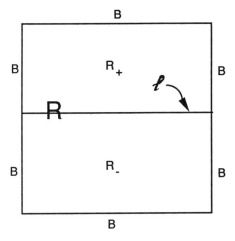

FIG. 7.5.

Proof. That the extension is harmonic (satisfies Laplace's equation) in R_- is easy to see. What is not clear is what happens *on* ℓ. To see this, let us consider the boundary values of u restricted to "upper" and "side" boundaries of R_+ (i.e., we exclude the points on $\{a < x < b \;, \; y = 0\}$. Extend those boundary values to the whole boundary B of the "big rectangle" R as an odd function of y. This will result in a continuous function f on B. Now solve the Dirichlet problem in R with boundary data f and call the resulting solution $U(x,y)$. We can show that U must be an odd function of y. To see this, consider the functions

$$v(x,y) = U(x,y) + U(x,-y).$$

v is harmonic in R and continuous in $\overline{R}$, and it vanishes on B, the boundary of R. By the uniqueness of the Dirichlet problem (or the maximum principle) $v \equiv 0$ in R. Hence U is odd in y. It follows that U vanishes on ℓ. Thus U has the same boundary values as u in $\overline{R}_+$. Since they are both harmonic in R_+ and continuous in $\overline{R}_+$, $u \equiv U$ in R_+. Thus U gives a harmonic extension of u as an *odd* function of y, as stated in the theorem.

A slight variant of the above proof gives the following result.

THEOREM 7.2.4. *Suppose we make the same assumptions as in Theorem 7.2.3, but instead of $u = 0$ on ℓ, we assume that the one-sided limit of $\partial u / \partial y$ vanishes on ℓ. Then u can be continued as a harmonic function to R by the following formula:*

$$u(x,y) = u(x,-y).$$

It is known that if a harmonic function u in a domain D vanishes in a nonempty open subset, then $u \equiv 0$ in D. This fact is called the *unique continuation* property. It implies that the harmonic extensions in Theorems 7.2.3 and 7.2.4 are unique.

From Theorems 7.2.3 and 7.2.4 we can deduce the following result.

THEOREM 7.2.5. *If u is harmonic in R_+ and the one-sided limits from above of u and $\partial u/\partial y$ vanish on ℓ, then $u \equiv 0$ in R_+.*

Indeed, by Theorems 7.2.3 and 7.2.4 both the odd and even extensions of u into R_- are harmonic extensions of u into R. Therefore also half their sum, which is zero in R_-, is harmonic extension of u into R_-. Thus the function

$$U(x,y) = \begin{cases} u(x,y) & \text{in} \quad R_+ \\ 0 & \text{in} \quad R_- \cup \ell \end{cases}$$

is harmonic in R. By the unique continuation property it follows that $u \equiv 0$ in R_+.

COROLLARY 7.2.1. *If two functions u_1 and u_2 are harmonic in R_+ and continuous in $\overline{R}_+$ and the one-sided limits agree, i.e.,*

$$u_1(x,y+0) = u_2(x,y+0), \qquad \frac{\partial u_1}{\partial y}(x,y+0) = \frac{\partial u_2}{\partial y}(x,y+0),$$

then $u_1 \equiv u_2$ in R_+.

Proof of Theorem 7.2.1. To make the proof more transparent we take $k = 1$. If D has the form (7.13), then the situation is as in Fig. 7.6.

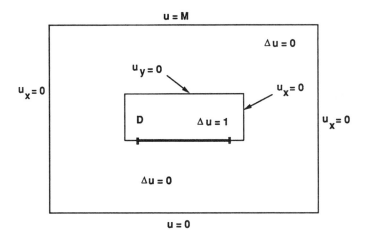

FIG. 7.6.

We reflect the rectangle D to the right (giving C), about its upper boundary (giving A), and reflect A to the right (giving B); see Fig. 7.7. B can also be obtained by reflecting C about its upper boundary. We next translate the coordinate system so that it is centered at the point where A, B, C, D meet.

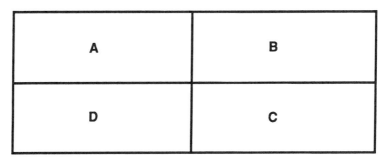

FIG. 7.7.

Thus B, A, D, and C are in the first, second, third, and fourth quadrants, respectively. By substracting $\frac{1}{2} x^2$ from u in D we get a harmonic function $\left(u - \frac{1}{2} x^2\right)$ in D whose x-derivative vanishes on the right boundary of D and thus has a unique harmonic extension to C given by

$$u(-x, y) - \frac{x^2}{2}, \quad (x, y) \in D, \quad (-x, y) \in C.$$

The limits from the right of this extension and of its x-derivative must agree with those of u on $x = 0$; hence the above extension must agree with the original function u in $C \cap R^+$; this follows from the corollary.

The same procedure yields a harmonic extension of $u(x, y) - \frac{1}{2} y^2$ in D to A, given by

$$u(x, -y) - \frac{y^2}{2}, \quad (x, y) \in D, \quad (x, -y) \in A,$$

and it agrees with the original u in $A \cap R^+$.

Now we can repeat the procedure in two ways. We can reflect the function given by the first formula from C to B about $y = 0$ to get a harmonic extension into B given by

$$(7.14) \qquad u(-x, -y) - \frac{x^2}{2}, \quad (x, y) \in D, \quad (-x, -y) \in B,$$

or we can reflect the function given by the second formula from A to B about $x = 0$ to get another harmonic extension into B given by

$$(7.15) \qquad u(-x, -y) - \frac{y^2}{2}, \quad (x, y) \in D, \quad (-x, -y) \in B.$$

By the same argument given for the first extension, each of the extensions constructed must agree with the original function u in B. However, since the two extensions into B differ, we get a contradiction.

PROBLEM 7.2.1. Does the proof of Theorem 7.2.1 extend to the case where D is a trapezoid?

Consider the situation just before the toner enters the development zone. Then

$$\Delta u = 0 \quad \text{in } R\backslash\{(x,0), -\varepsilon \le x \le \varepsilon\}$$

and u satisfies (7.8) and the boundary conditions (7.10). We can show that if $M < \sigma_* h$ and $h < b$ then

(7.16)
$$\frac{\partial u}{\partial y}(x, 0+) < 0 \quad \text{if } -\varepsilon < x < \varepsilon,$$

(7.17)
$$\frac{\partial^2 u}{\partial x \partial y}(x, 0+) < 0 \quad \text{if } 0 < x < \varepsilon.$$

In view of the explanation in (7.3), the inequality (7.16) means "move out the interface," that is, a toner layer will accumulate on $y = 0$ for $-\varepsilon \le x \le \varepsilon$. The rate at which this accumulation occurs is determined by the size of $n \cdot \vec{E}_{\text{air}}$. This quantity, according to (7.17), increases in absolute value as x increase from 0 to ε. Thus we expect that for small time the partially developed visible image will have the form depicted in Fig. 7.8.

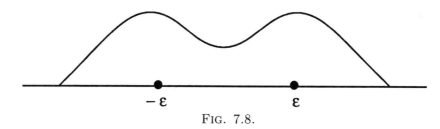

$-\varepsilon$ ε

FIG. 7.8.

Near the edges $x = \pm\varepsilon$ the dark image is more pronounced than in the center. This, in fact, is a well-known phenomenon in xerography; it is called the *edge enhancement* effect.

PROBLEM 7.2.2. Establish (7.16) numerically.

PROBLEM 7.2.3. Suppose D is part of a circle that contains the points $(\pm\varepsilon, 0)$ and a point $(0, \ell)$ on the y-axis. Compute the solution of (7.6)–(7.10) for this D with small ℓ, and find the sign of $\partial u/\partial n$ on Γ. Check whether $\vec{n} \cdot \vec{E}_{\text{air}} < 0$ on all of Γ.

7.3. Summary

In this section we focused on the process that converts the electric image into the visible image. After describing the physical process, we considered the mathematical model, which is based on the notion of a "free boundary problem," i.e., the problem of finding the region containing the toner. It was shown that in the ideal situation where the electric image is a rectangle, a certain amount of distortion in the visible image is unavoidable, i.e., the visible image cannot be a rectangle. The reflection principle and other properties of harmonic functions are the tools used in proving this result. Exercises explored other properties of the "free boundary."

References

An introduction to electrophotography is given in references [1], [2] of the previous chapter.

The model presented in this chapter was presented by John Spence in June 1989 and is described in

[1] A. Friedman, *Mathematics in Industrial Problems, Part* 2, Chap. 17, IMA Vol. Math. Appl., Vol. 24, Springer-Verlag, New York, 1989.

Existence of solution to the stationary problem (7.6)–(7.11) was proved under some restrictions on ε (ε "near" 0 or ε "near" a) in

[2] A. Friedman and B. Hu, *A free boundary problem arising in electrophotography*, Nonlinear Anal., 9 (1991), pp. 729–759.

[3] B. Hu and L. Wang, *A free boundary problem arising in electrophotography: solutions with connected toner region*, SIAM J. Math. Anal., 23 (1992), pp. 1439–1454.

The time dependent problem was studied in

[4] A. Friedman and J. J. L. Velázquez, *A time-dependent free boundary problem modeling the visual image in electrophotography*, Archive Rat. Mech. Anal., 123 (1993), pp. 259–303.

In this paper a unique solution is proved to exist for small time, and the edge enhancement phenomenon is established.

Index

135